Ground Water Modeling in Multilayer Aquifers

Unsteady Flow

M. M. Aral

LEWIS PUBLISHERS

Library of Congress Cataloging-in-Publication Data

Aral, M. M. (Mustafa M.)
Ground water modeling in multilayer aquifers : unsteady flow

Includes bibliographical references. (p.).
1. Water, Underground—Mathematical models. 2. Water, Underground—Data processing. 3. Water-supply—Mathematical models. 4. Water-supply—Data processing. 5. Unsteady flow (Fluid dynamics)—Mathematical models. 6. Unsteady flow (Fluid dynamics)—Data processing. I. Title.
GB1001.72.M35A74 1990 551.49'01'5118 89-13719
ISBN 0-87371-305-2

While the computer code supplied with this book has been tested and verified by the author over a period of five years, no responsibility is assumed by the author or the publisher for any errors, mistakes, or misrepresentations that may occur from the use of these programs. No warranty of any kind, expressed or implied, is made by the author, including, but not limited to, the implied warranties of merchantability and fitness for a particular purpose and to the accuracy and functioning of the program and the related program material. The author does not assume any responsibility and liability for adverse conditions resulting from misuse and misinterpretation of the computer code ULAM.EXE and related program material. The author shall not be liable for errors contained herein or for incidental or consequential damages in connection with the furnishing, performance, or use of this material and the computer code accompanying this text. The information in this document and the computer code supplied with the distribution disk is subject to change without notice.

IBM PC/AT, PS/2 are trademarks of IBM Corporation.
Microsoft FORTRAN, MS-DOS are trademarks of Microsoft Corporation.
Golden Graphics System is a trademark of Golden Software Inc.

Computer Code : ULAM.FOR, ULAM.EXE

LEWIS PUBLISHERS, INC.
121 South Main Street, Chelsea, Michigan 48118

PRINTED IN THE UNITED STATES OF AMERICA

Preface

In this book and the enclosed software a computational management tool designed for use by hydrogeologists, engineers, geologists, educators, and students is presented. The software is suitable for implementation on microcomputers for the analysis of unsteady flow in multilayer aquifers. Unsteady flow in a typical multilayer aquifer system may be generated in response to time-dependent pumping operations from well networks distributed in the multilayer aquifer system or may be developed within the system in response to time-dependent aquifer boundary conditions.

The multilayer aquifer computer program discussed in this text and included in the distribution disk is called the Unsteady Layered Aquifer Model (ULAM.EXE). The distribution disk accompanying this text also contains several data files for case studies discussed in the text and a BASIC program (IDEAL.BAS), which may be used to plot the finite-element idealization of the aquifer under study. Procedures to implement these codes are explained in detail in the text. Throughout the text, emphasis has been placed on practice and applications rather than on theory and mathematical rigidity.

Though this book was designed primarily to supplement the software included in the distribution disk, the basic physical and mathematical concepts explained in it will also be useful in clarifying various aspects of regional ground water flow studies and application of finite-element analysis to multilayer aquifer problems. It includes basic theory, assumptions involved, and restrictions and limits of the model presented for unsteady multilayer aquifer analysis, along with the finite-element formulation of the governing partial differential equations. Although all necessary information is included in this book, the reader is expected to have a working knowledge of the basic concepts of geohydrology to fully implement the computer code.

The computer program for unsteady flow analysis is easy to use and as general as possible, but because of the complexity of multilayer aquifer applications that may be generated and analyzed with this software, input data preparation can involve a great deal of work. To ease the task of data preparation, data generation routines are included in the computer code. A data file input format, rather than a menu-operated *screen* input format, is used to facilitate problem definition and accuracy of data input. Output from the computer code is arranged in an easy-to-follow format and is stored in output data files. These output data files may later be printed out, viewed on the screen, or utilized as input data files for commercially available plotting routines for contour plotting or surface plotting of numerical results, as demonstrated in this book.

The input data preparation format presented in this text basically follows the procedures described in the earlier text on steady-state multilayer aquifer flow (Aral 1990), with a few additions that characterize the time-dependent nature of unsteady flow problems. The data generation routines are similar to the ones used in the steady-state model, and, in fact, the data prepared for the steady-state model may be extended to an unsteady-state data base with minimal editing. For those who have not worked with finite-element applications in the past, I highly recommend the use of the steady-state model first.

This book is an outgrowth of my lectures to graduate students in the School of Civil Engineering at Georgia Institute of Technology and short courses for consulting and industry personnel. I am grateful for the comments and suggestions received from my colleagues and students who reviewed and studied the material from which this book evolved. In particular, I would like to thank Elie Haddad, who prepared parts of the earlier versions of the code and verified the code using several test problems, and my son, Sinan, for his patience in editing this book and his continuous support during this project.

Mustafa M. Aral
Atlanta, Georgia

Mustafa M. Aral is a faculty member at the School of Civil Engineering, Georgia Institute of Technology, Atlanta, Georgia, specializing in geohydrology. He received his BS in Civil Engineering from Middle East Technical University, Ankara, Turkey in 1967 and his MS and PhD from Georgia Institute of Technology in 1969 and 1971, respectively.

Dr. Aral is the author of three books and over 50 technical papers in the field of fluid mechanics, ground water hydrology, and computational methods. He is actively involved in consulting with government agencies, as well as consulting firms specializing in the general area of geohydrology, and has lectured at many universities throughout the United States, Canada, Europe, and the Middle East.

Contents

List of Figures

List of Tables

Ground Water Modeling in Multilayer Aquifers

Unsteady Flow

1

Introduction

In geologic formations, whether externally disturbed or not, water is continuously in a state of motion. More often than not, these geologic formations are layered, nonhomogeneous, and anisotropic, requiring a three-dimensional representation and analysis. In these geologic systems, quantification of the volume rate of ground water flow and the determination of local pressures, direction of flow, and its velocity are extremely difficult tasks. However, determination of these parameters is an essential part of all ground water quality and quantity analyses that need to be routinely performed in most environmental and hydrogeologic studies. The problem of analysis of unsteady ground water flow in a heterogeneous aquifer system has stimulated great interest, has presented significant mathematical difficulties, and has been of growing importance in ground water management applications over the last two decades.

The solution and analysis of three-dimensional problems by analytical methods is extremely complex unless one idealizes the aquifer system as symmetric, homogeneous, and isotropic, but in such an approach, the solution obtained loses its practical applicability, since most aquifers are nonsymmetric, heterogeneous, and anisotropic. Numerical solution is the other common alternative used in typical application studies. Even in this more friendly environment, the solution of ground water flow problems is again performed, in most instances, in two dimensions to avoid the complexity and high cost of three-dimensional numerical modeling. However, in cases such as multilayer systems in which there are hydraulic connections between the alternating aquifer layers or in which aquifer properties show significant variation in the third dimension, or both, two-dimensional numerical solutions are not justifiable. In this book these systems are identified as multilayer aquifer systems.

In multilayer systems, main aquifers, whether semiconfined or unconfined, may lose or gain water through either or both of the geologic formations bounding the main aquifer layers. Although the confining geologic formations in these systems usually have relatively high resistance to flow, leakage between the semiconfined aquifers cannot be completely ignored, because large horizontal contact areas are involved and large piezometric head differences may exist between the main aquifer layers. Pressure difference is the principal driving force for leakage, and the storage effect of the aquitards and aquifers is the main cause of delayed response of ground water movement, points which should be considered in all unsteady flow analysis.

In the past, these systems were frequently analyzed using simplified models. The complexity of these models ranges from one-dimensional analytical solutions (Bear 1979) to axisymmetric applications (Neuman and Witherspoon 1969a, 1969b) to numerical solutions that simulate complex regional semiconfined or confined main aquifer layers (Pinder and Gray 1977). When microcomputers are to be used, most three-dimensional formulations are out of reach, primarily because of the constraints on available computer memory and the tasking CPU processing time required. Primary objectives with this book are to present a mathematical–numerical model and a computer program that will analyze unsteady flow in a multilayer aquifer system as realistically as possible without complicating the solution processes *and* to present the code in a form suitable for implementation on a microcomputer.

One can usually simplify multilayer aquifer problems by assuming that the vertical components of flow within semiconfined and unconfined aquifers are negligible and that only the components of flow in the plane of the aquifer need to be considered. To simplify the multilayer ground water seepage problem further, one can assume that the horizontal components of flow in the semiconfining layers are negligible and that only the vertical components need to be considered. Furthermore, for unsteady-state problems, the time-dependent rate of leakage through a confining bed may be defined in terms of piezometric head values of the adjacent main aquifer layers and the aquitard's hydraulic conductivity and specific storage coefficient. These basic assumptions are generally valid for a field situation in which the difference in hydraulic conductivity between the main aquifers and the confining layers is at least two orders of magnitude (Hantush 1960; Bredehoeft and Pinder 1970).

Given these assumptions, the three-dimensional flow problem in a multilayer aquifer system can be reduced to a quasi three-dimensional form in which the two-dimensional aquifer solutions in the plane of the aquifer are linked vertically through leakage terms. Mathematically, this simplification reduces the three-dimensional equations governing flow in a multilayer aquifer system to the solution of coupled partial differential equations that are two-dimensional in the plane of the aquifer.

In this book these equations are solved utilizing finite-element Galerkin procedures. The multilayer aquifer system may be a fully confined system or may have an unconfined upper layer that is hydraulically connected to the semiconfined layers below. The computer code generates internally the multilayer finite-element data and the governing coupled matrix equations using a reference base layer finite-element idealization, thereby reducing the data preparation phase considerably. The governing matrix equations are then solved by standard techniques, yielding piezometric head distribution in each layer simultaneously for all time periods. Field parameters for the main and confining layers should be supplied externally using the data input file format described in detail in Chapter 4.

The computer model is capable of analyzing nonhomogeneous, anisotropic multilayer aquifer systems that may or may not have a free surface aquifer as its uppermost layer. Any of these layers may be pumped or recharged as a function of time through wells distributed in the aquifer system. The model may also treat heterogeneous and time-dependent infiltration conditions if an unconfined aquifer is modeled. The computer code may be used in the analysis of a multilayer aquifer system that has up to five main and six semiconfined aquifer layers, or, if desired, a simple, one-layer fully confined or unconfined regional aquifer study may be performed. Case studies are included to demonstrate the versatility of the computer code and the limits of its application. The simplicity of the numerical procedures and the flexibility of the data generation routines will make the technique and the computer code immediately applicable to practical field situations.

A word of caution is always necessary in all modeling efforts. Model development is an iterative process between the conceptual progress and actual understanding of the physical phenomena modeled. One usually starts with preliminary ideas and develops the first conceptual model. Through experience, evaluation, analysis, and interpretation of the physical system, ideas are revised and a better conceptual model is developed. This leads to a second model. This iterative process should always be followed in every modeling effort. In this process some steps may have to be repeated, and in some cases even the starting point and initial assumptions should be challenged. At the end, one should also realize that the final product will always be an approximation of the physical system modeled and not the actual system. Within this perspective, the software presented in this book is recommended as a practical computational tool.

2

Mathematical Model

In this chapter the mathematical model for unsteady ground water flow analysis in a multilayer aquifer system is presented in a conventional way. In developing numerical methods for the analysis of field problems in ground water flow, it is customary to start by defining the governing equations in the form of partial differential equations. These equations are the mathematical representation of the physical phenomena occurring at a point in the flow domain. The next step is the analytical or numerical integration of these equations in space and time to define exact or approximate solutions.

When heterogeneity and anisotropy of aquifers are considered, the partial differential equation describing unsteady flow in a typical aquifer system needs to be developed in a three-dimensional domain. The difficulty in solving engineering applications, defined by these partial differential equations, grows exponentially when one recognizes the heterogeneity of the natural boundary conditions in the three-dimensional domain, which needs to be satisfied; the chronic lack of field data to represent characteristic field parameters of the aquifer in three dimensions; and the nonlinearity of the governing partial differential equations when unconfined aquifers are studied. To overcome these difficulties, ground water flow problems are usually categorized into simpler forms, such as *confined*, *unconfined*, and *leaky* aquifers, and closed form solutions are sought for these idealized situations. A more realistic analysis may also be made for these idealized aquifers if numerical methods are employed, in which case heterogeneity and anisotropy of field parameters may be introduced into the analysis. In this book, a quasi three-dimensional transient model is developed for the analysis of a multilayer aquifer system, and since numerical procedures are employed in the solution, the model generated may be employed in the analysis of regional heterogeneous and anisotropic aquifer systems of practical interest.

GOVERNING EQUATIONS IN MAIN AQUIFER LAYERS

Consider a regional aquifer system, for which a typical definition sketch is illustrated in Figure 1. Configuration of each layer can be any one of the cross sections shown in Figures 2–4, i.e., a confined flow, a semiconfined flow, an unconfined flow, or any combination. The derivation of the governing equations for such a system is well established in the literature (Bear 1979; Huyakorn and Pinder 1983; Strack 1989). Hence, it is sufficient to relate only to the underlying assumptions and resulting equations governing flow in a multilayer aquifer system.

The conceptual basis for the transient fluid flow equation is the law of conservation of mass and Darcy's law. Darcy's law furnishes three equations of motion for the four primary unknowns in a typical ground water study. These unknowns are the three specific discharge vectors, q_x, q_y, q_z, and the piezometric head h. For an anisotropic hydraulic conductivity, where the principal hydraulic conductivity directions are parallel to the main coordinate directions chosen, these equations take the form:

$$q_x = -K_x \frac{\partial h}{\partial x}; \qquad q_y = -K_y \frac{\partial h}{\partial y}; \qquad q_z = -K_z \frac{\partial h}{\partial z} \tag{2.1}$$

For the case where the principal hydraulic conductivity directions are *not* parallel to the main coordinate directions, the coefficient of permeability should be defined as a tensor (see Appendix E). When numerical methods are employed in the solution, such cases may always be reduced to the form given in Equation

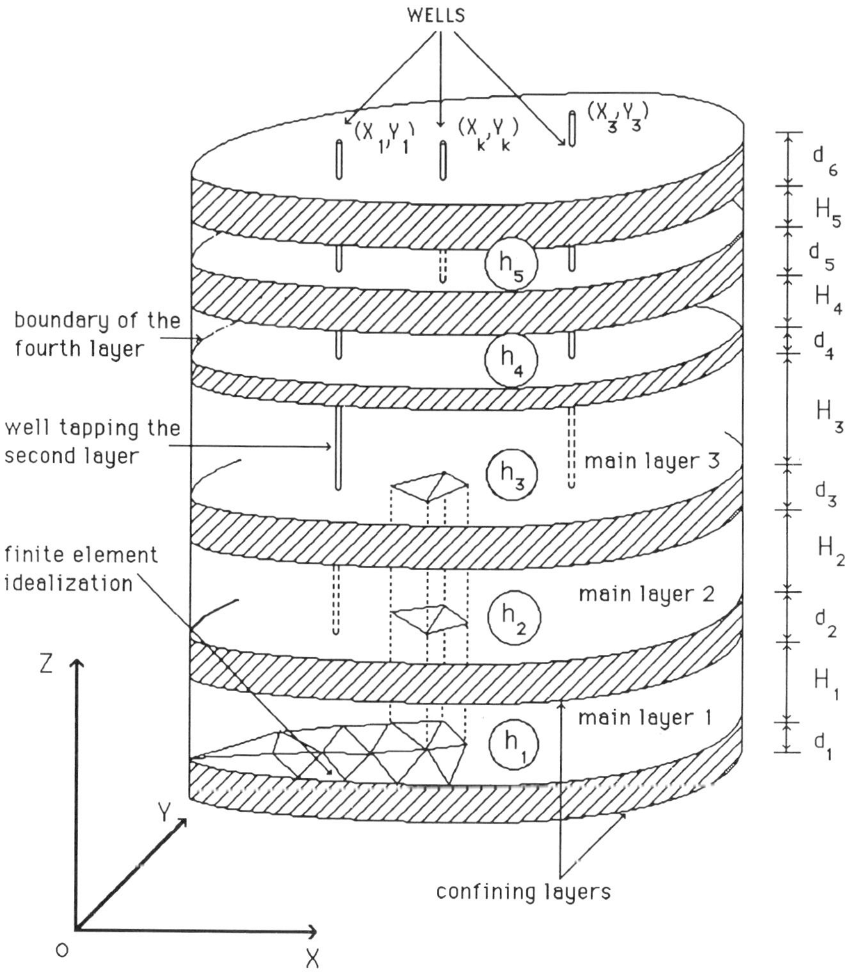

Figure 1. Definition sketch for a multilayer aquifer system.

2.1 by elementary coordinate transformations. These coordinate transformations are treated in more detail when the specifics of the numerical methods are discussed in Chapter 3.

A fourth equation may be obtained from the mass balance

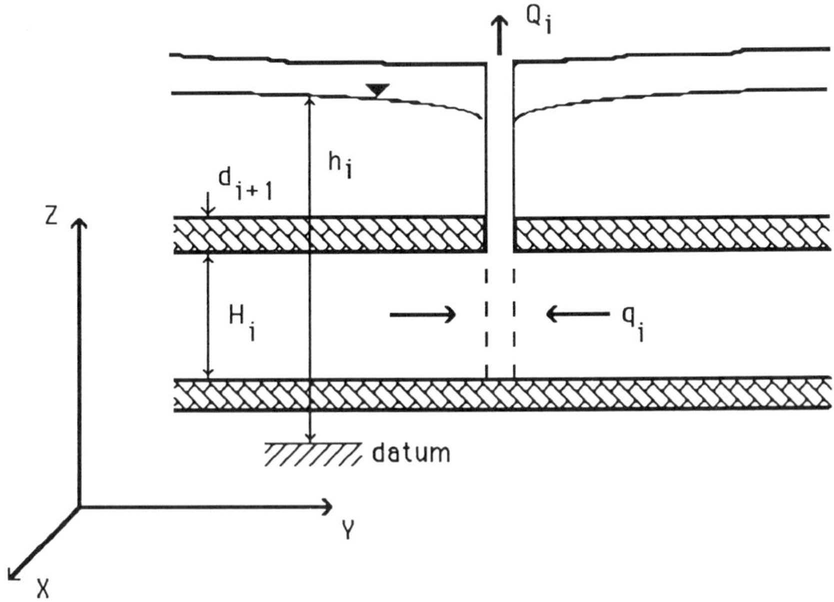

Figure 2. Definition sketch for vertical cross section of a confined aquifer.

analysis by writing the inflow-outflow balance around an infinitesimal control volume that extends vertically from the top to the bottom of the aquifer. For a fully saturated unsteady flow problem, the resulting equation takes the form:

$$\frac{\partial q_x}{\partial x} + \frac{\partial q_y}{\partial y} + \frac{\partial q_z}{\partial z} + S_s \frac{\partial h}{\partial t} = 0 \tag{2.2}$$

or assuming that Darcy's law is valid, one may write Equation 2.2 as:

$$\frac{\partial}{\partial x}(K_x \frac{\partial h}{\partial x}) + \frac{\partial}{\partial y}(K_y \frac{\partial h}{\partial y}) + \frac{\partial}{\partial z}(K_z \frac{\partial h}{\partial z}) = S_s \frac{\partial h}{\partial t} \tag{2.3}$$

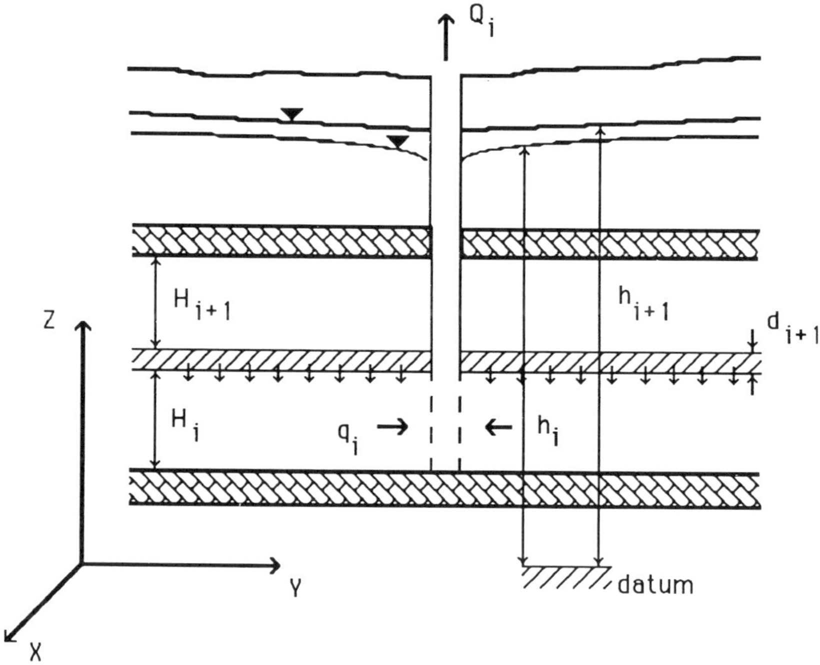

Figure 3. Definition sketch for vertical cross section of a two-layer aquifer system.

which is the governing equation for unsteady three-dimensional ground water flow in an anisotropic porous medium. Solving unsteady-state ground water flow problems amounts to solving this partial differential equation in space and time with appropriate choice of boundary and initial conditions. Boundary conditions will be given in terms of (1) the piezometric head h, which will be identified as a Dirichlet boundary condition; (2) a discharge vector, which will be identified as a Neumann boundary condition; or (3) a head-dependent discharge vector, which will be identified as a Cauchy or head-dependent boundary condition.

In order to generate a simpler mathematical model for multi-

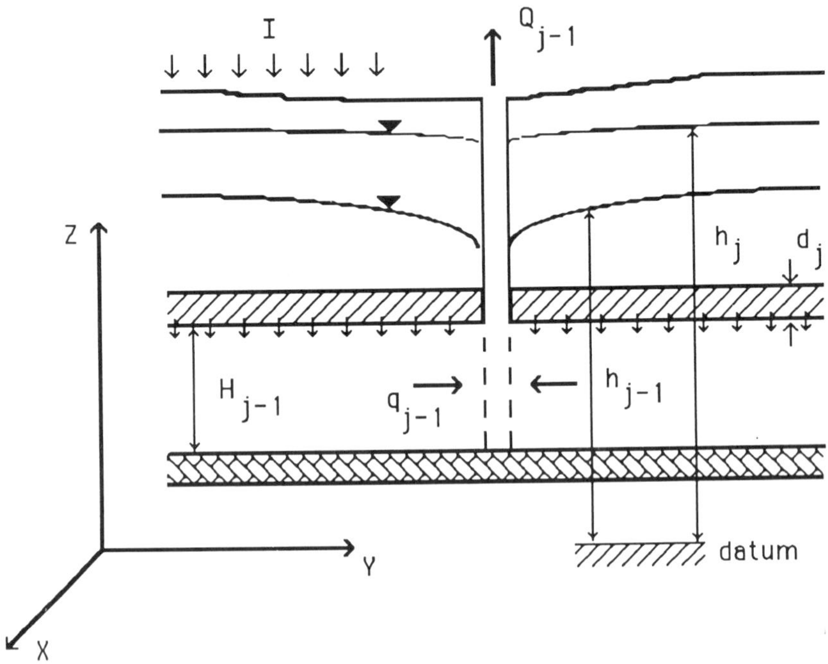

Figure 4. Definition sketch for vertical cross section of a two-layer semiconfined-unconfined aquifer combination.

layer aquifers, the equation given above can be reduced to a quasi three-dimensional form with the following assumptions:

Assumption 1. The flow in the main aquifer layers is two-dimensional in the horizontal plane of the aquifer. While the aquifer thickness may vary slightly, this variation is much smaller than the horizontal dimensions under consideration; thus, the flow in the main aquifer layers may be considered nearly vertically uniform.

Assumption 2. The direction of flow is vertical in the confining layers (aquitards).

Assumption 3. Main aquifer layers are hydraulically connected through leakage from confining layers, but the difference between the hydraulic conductivity of the main aquifers and the confining layers is large.

Assumption 4. The flow is unsteady in a fully saturated porous medium, and Darcy's law applies.

Assumption 5. Aquifers and aquitards are generally nonhomogeneous, and aquifers are anisotropic.

Assumption 6. Storage effects of aquifers as well as aquitards are important in defining the flow conditions in the multilayer aquifer system.

Given these assumptions, the partial differential equation given above may be integrated in the vertical direction, yielding a vertically averaged two-dimensional form of the governing equation (2.3). Mathematical details of this process are summarized in Appendix C. For the ith semiconfined aquifer layer tapped by several wells, this equation may be reduced to the form given below:

$$S_{si} H_i \frac{\partial h_i}{\partial t} - \frac{\partial}{\partial x}(T_{xi} \frac{\partial h_i}{\partial x}) - \frac{\partial}{\partial y}(T_{yi} \frac{\partial h_i}{\partial y}) + q_{zi} \Big|_{z_2} - q_{zi} \Big|_{z_1}$$

$$- \sum_{k=1}^{w} Q_i(x_k, y_k) = 0 \qquad i = 1, 2, 3, ..., J \qquad (2.4)$$

where J is the number of aquifer layers. Under the same assumptions, Equation 2.4 takes the form shown below for an unconfined top aquifer layer that is tapped by several wells, with infiltration as shown in Figure 4:

$$S_{yj} \frac{\partial h_j}{\partial t} - \frac{\partial}{\partial x}(K_{xj} h_j \frac{\partial h_j}{\partial x}) - \frac{\partial}{\partial y}(K_{yj} h_j \frac{\partial h_j}{\partial y}) - q_{zj} \Big|_{z_1} + I(x,y)$$

$$- \sum_{k=1}^{w} Q_j(x_k, y_k) = 0 \qquad (2.5)$$

where $h_i(x,y)$ is the vertically averaged piezometric head in an aquifer layer i; $T_{xi}(x,y)$ and $T_{yi}(x,y)$ are the transmissivity values of the main aquifer layers; $K_{xj}(x,y)$ and $K_{yj}(x,y)$ are the hydraulic conductivities of the top unconfined aquifer layer in x and y directions, respectively; Q_i is the volumetric well discharge or recharge from the ith aquifer layer at location (x_k,y_k); w is the total number of wells in the ith aquifer; I(x,y) is the infiltration or evaporation rate, which is only defined for an unconfined aquifer layer; S_s is the specific storage ($S = S_sH$ is the storage coefficient); S_y is the specific yield defined for unconfined aquifers (sometimes called effective porosity). The leakage terms, q_{zi}, will be defined in terms of the analytical solution of the governing equations in the aquitards and Darcy's law as shown below. It is further assumed that either the confining layers 1 and (J + 1) are impervious, i.e., the J-layer system is fully confined, or the piezometric head distributions h_o and h_j are known, and only h_j may vary in response to flow situations in the main aquifers below it, simulating an unconfined aquifer layer. As can be seen in Equation 2.5, the equation given for the unconfined aquifer is nonlinear. When a numerical solution is attempted, this equation will be linearized and the solution will be obtained iteratively as described in Chapter 3.

THE GOVERNING EQUATION IN AQUITARDS AND THE ANALYTICAL SOLUTION

Since the flow is assumed to be unidirectional in the aquitards, the governing equation describing the piezometric head distribution in an aquitard can be given as follows:

$$K_{za_i} \frac{\partial^2 h_{a_i}}{\partial z^2} = S_{sa_i} \frac{\partial h_{a_i}}{\partial t} \qquad i = 1, 2, 3, \ldots, (J+1) \qquad (2.6)$$

where h_{ai} represents the piezometric head in the ith aquitard, z is the vertical coordinate, S_{sai} is the specific storage of the ith aquitard, and K_{zai} is the hydraulic conductivity of the ith aquitard in the z coordinate direction. Equation 2.6 may now be solved analytically as a function of time and space in terms of the piezometric head variables h_i and h_{i-1}, which represent the time-dependent vertically averaged piezometric head values in the adjacent main aquifer layers (Figure 5).

Let us now define $\bar{h}_i$, h_i as the piezometric head values in the ith aquifer at time $(t + \Delta t)$ and (t), respectively, and $\bar{h}_{i-1}$, h_{i-1} as the piezometric head values in the $(i - 1)^{th}$ aquifer at time $(t + \Delta t)$ and (t), respectively. These terms represent the boundary conditions of h_{ai} at $(z = 0, L)$ at the beginning and end of the solution time interval, which may be identified as Δt. Using these definitions, the analytical solution to Equation 2.6 can be given as a function of z and t as follows (Aral and Haddad 1989):

$$h_{a_i}(z,t) = (\bar{h}_i - \bar{h}_{i-1})(z / L_i) + \bar{h}_{i-1}$$

$$+ [(2L_i\delta_i + 4\beta_i) / \pi] \sum_{n=1}^{\infty} \{[1/(2n-1)] \exp(-(2n-1)^2\alpha_i^2 \pi^2 t / L_i^2) \times \sin[(2n-1)\pi z / L_i]\}$$

$$- [2L_i\delta_i / \pi] \sum_{n=1}^{\infty} \{(1/2n) \exp(-4n^2 \alpha_i^2 \pi^2 t / L_i^2) \times \sin(2n\pi z / L_i)\} \quad (2.7)$$

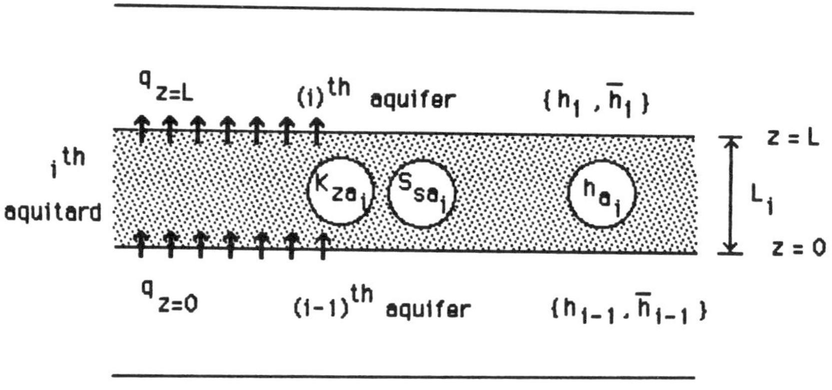

Figure 5. Definition sketch for aquitard leakage.

in which:

$$\delta_i = [(h_i - \bar{h}_i) - (h_{i-1} - \bar{h}_{i-1})] / L_i \tag{2.8}$$

$$\beta_i = (h_{i-1} - \bar{h}_{i-1}) \tag{2.9}$$

$$\alpha_i^2 = K_{za_i} / S_{sa_i} \tag{2.10}$$

and L_i is defined as the thickness of the ith aquitard at a given location in the multilayer aquifer system where the analytical solution is defined.

The solution given in Equation 2.7 represents the variation of the piezometric head variable in the aquitard in the vertical direction as a function of z and t. Using this analytical solution and Darcy's law, leakage terms can now be computed at the top and the bottom of the ith aquitard at any time, as shown below. These leakage terms will be a function of the aquitard properties and the piezometric head variables of main aquifer layers which are above and below the ith aquitard. When one substitutes the resultant leakage equations into Equation 2.4 or 2.5, the primary unknown in these coupled partial differential equations will be the vertically averaged piezometric head variables of the main aquifer layers. These equations may then be solved numerically using standard finite-element Galerkin procedures for these primary unknowns (h_i).

ANALYTICAL REPRESENTATION OF LEAKAGE TERMS

Given the analytical solution for the piezometric head distribution in the ith aquitard (Equation 2.7), the leakage terms at the bottom and top of the aquitard may now be computed. From Darcy's law we know that:

$$q_{z_i}(z,t) = -K_{za_i} \frac{dh_{a_i}}{dz} \tag{2.11}$$

Evaluating the derivative of the piezometric head from Equation 2.7, one may write Equation 2.11 as:

$$q_{z_i} = K_{za_i} (\bar{h}_{i-1} - \bar{h}_i) / L_i$$

$$- [K_{za_i}(2L_i\delta_i + 4\beta_i) / L_i] \sum_{n=1}^{\infty} \{\exp[(-2n - 1)^2 \alpha_i^2 \pi^2 t / L_i^2] \times \cos[(2n - 1) \pi z / L_i]\}$$

$$+ 2K_{za_i}\delta_i \sum_{n=1}^{\infty} \{\exp(-4n^2 \alpha_i^2 \pi^2 t / L_i) \cdot \cos(2n\pi z / L_i)\} \quad (2.12)$$

Equation 2.12 may now be used to evaluate the leakage terms at the bottom and the top of the ith aquitard, which will be considered leakage into or out of the main aquifer layers (i) and (i – 1) from the bottom and the top of the aquifer layers, respectively (Figure 5).

At $z = 0$:

$$q_{z_i}(0,t) = K_{za_i} (\bar{h}_{i-1} - \bar{h}_i) / L_i$$

$$- [K_{za_i}(2L_i\delta_i + 4\beta_i) / L_i] \sum_{n=1}^{\infty} \exp[-(2n - 1)^2 \alpha_i^2 \pi^2 t / L_i^2]$$

$$+ 2K_{za_i}\delta_i \sum_{n=1}^{\infty} \exp[-4 n^2 \alpha_i^2 \pi^2 t / L_i^2] \quad (2.13)$$

and at $z = L$:

$$q_{z_i}(L,t) = K_{za_i} (\bar{h}_{i-1} - \bar{h}_i) / L_i$$

$$+ [K_{za_i}(2L_i\delta_i + 4\beta_i) / L_i] \sum_{n=1}^{\infty} \exp[-(2n - 1)^2 \alpha_i^2 \pi^2 t / L_i^2]$$

$$+ 2K_{za_i}\delta_i \sum_{n=1}^{\infty} \exp[-4 n^2 \alpha_i^2 \pi^2 t / L_i^2] \quad (2.14)$$

and the definitions of δ and β are as given in Equations 2.8 and 2.9. Leakage terms may now be computed from Equations 2.13 and 2.14 and introduced into Equation 2.4 or 2.5. For an ith *aquifer layer,* the leakage terms may be given as Equation 2.15.

$$
\begin{aligned}
\bar{q}_i &= q_{zi}\big|_{z_2} - q_{zi}\big|_{z_1} \\
&= \bar{h}_i \left[\frac{K_{za_{i+1}}}{L_{i+1}} (1 + 2\{SUM1_{i+1}\} + 2\{SUM2_{i+1}\}) \right. \\
&\qquad \left. + \frac{K_{za_i}}{L_i} (1 + 2\{SUM1_i\} + 2\{SUM2_i\}) \right] \\
&\quad - h_i \left[\frac{2K_{za_{i+1}}}{L_{i+1}} (\{SUM1_{i+1}\} + \{SUM2_{i+1}\}) \right. \\
&\qquad \left. + \frac{2K_{za_i}}{L_i} (\{SUM1_i\} + \{SUM2_i\}) \right] \\
&\quad - \bar{h}_{i+1} \frac{K_{za_{i+1}}}{L_{i+1}} (1 - 2\{SUM1_{i+1}\} + 2\{SUM2_{i+1}\}) \\
&\quad - h_{i+1} \frac{2K_{za_{i+1}}}{L_{i+1}} (\{SUM1_{i+1}\} - \{SUM2_{i+1}\}) \\
&\quad - \bar{h}_{i-1} \frac{K_{za_i}}{L_i} (1 - 2\{SUM1_i\} + 2\{SUM2_i\}) \\
&\quad - h_{i-1} \frac{2K_{za_i}}{L_i} (\{SUM1_i\} - \{SUM2_i\})
\end{aligned}
\tag{2.15}
$$

where

$$\{SUM1_i\} = \sum_{n=1}^{\infty} \exp\left[-(2n - 1)^2 \alpha_i^2 \pi^2 t / L_i^2\right] \tag{2.16}$$

and

$$\{SUM2_i\} = \sum_{n=1}^{\infty} \exp\,[-4\; n^2\; \alpha_i^2\; \pi^2\; t\; /\; L_i^2] \tag{2.17}$$

As is the case in steady flow analysis, the resultant leakage terms are now functions of h_i, h_{i-1}, and h_{i+1}. But this relationship is now time-dependent and has to be computed in terms of the piezometric head variables at times $(t + \Delta t)$ and (t). Nevertheless, when Equation 2.15 is substituted into Equation 2.4 or 2.5, the resulting equations governing flow in main aquifer layers will be in terms of piezometric head variables of the main aquifer layers. Standard finite-element Galerkin procedures may now be employed in numerical solution of the governing equations for the unknowns, h_i.

BOUNDARY CONDITIONS

Ground water flow in the main aquifer layers is governed by conditions at the boundaries of the regional system; the piezometric head distributions h_o and h_j; the infiltration, if an unconfined aquifer is specified; and the recharge and/or pumping well distributions in the respective aquifers. In a typical aquifer study, three types of boundary conditions may be defined. Each of these boundary conditions may act on a segment of the overall boundary of the ith aquifer layer (B_{1i}, B_{2i}, B_{3i}), as shown in Figure 6. These boundary conditions may be defined as follows:

Dirichlet Boundary Condition

On boundary segment B_{1i}, a condition of prescribed piezometric head, or the so-called Dirichlet condition, may be defined as:

$$h_i(x,y,t) = g_i(x,y,t) \qquad \text{for } (x,y) \text{ on } B_{1i} \tag{2.18}$$

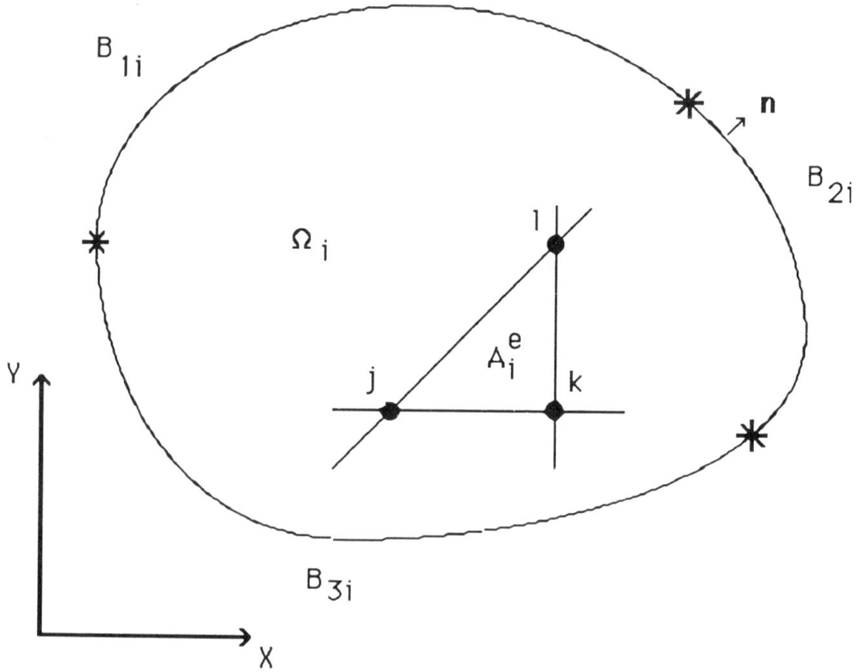

Figure 6. Aquifer domain, boundaries, and a typical element.

where $g_i(x,y,t)$ is a known function. Although this function is described as a continuous function above, it is understood that as input data, only discrete values of this function at the boundary nodes at certain time intervals are needed. This boundary condition may occur at the interface of the saturated porous medium and an open body of water such as the bank of a canal, lake, or river. This type of a boundary condition may also be used to describe conditions at a far field boundary, i.e., at a boundary segment where no appreciable change in the piezometric head value is expected.

Neumann Boundary Condition

The boundary of prescribed flux, that is, a Neumann condition, may be defined on another segment of the boundary identified as B_{2i} (Figure 6). With **n** defined as the outward unit normal vector on B_{2i}, the mathematical statement of this boundary condition can be given as:

$$(\mathbf{T}_i \cdot \nabla h_i) \cdot \mathbf{n} = q_i(x,y,t) \qquad \text{for } (x,y) \text{ on } B_{2i} \qquad (2.19)$$

in which $q_i(x,y,t)$ is a known function defined at the boundary B_{2i} of the ith layer of the regional system, $\mathbf{T}_i$ is the transmissivity tensor with principal components T_{xi} and T_{yi}, and ∇ is the two-dimensional gradient operator. A special form of this boundary condition frequently occurs in ground water flow studies. For impervious boundaries or for boundaries where a water divide is specified, the flux across the boundary needs to be defined as zero, in which case the function $q_i(x,y,t)$ assumes a zero value.

Head-Dependent Boundary Condition

The Head-dependent (Cauchy) boundary condition may be identified as induced infiltration and can be illustrated by an example. Consider the boundary B_{3i} in Figure 6 to represent a line source such as a river bed, for which a typical cross section is shown in Figure 7. The leakage per unit length into the aquifer, q_r, may be given by:

$$(\mathbf{T}_i \cdot \nabla h_i) \cdot \mathbf{n} = (K_r\, w_r / m_r)(h_r - h_i) = q_r \qquad \text{for } (x,y) \text{ on } B_{3i} \quad (2.20)$$

where K_r is the permeability, m_r is the thickness, and w_r is the width of the river bed, respectively, and h_r is the elevation of the river stage, which may be time-dependent. When h_i is less than z_r, the elevation of the river bottom, q_r, reaches a maximum, and thus a more complete statement of this boundary condition may be given as follows:

$$q_r = \begin{cases} (K_r\, w_r / m_r)(h_r - h_i) & \text{for } h_i > z_r \\ (K_r\, w_r / m_r)(h_r - z_r) & \text{for } h_i \le z_r \end{cases} \qquad \text{for } (x,y) \text{ on } B_{3i} \quad (2.21)$$

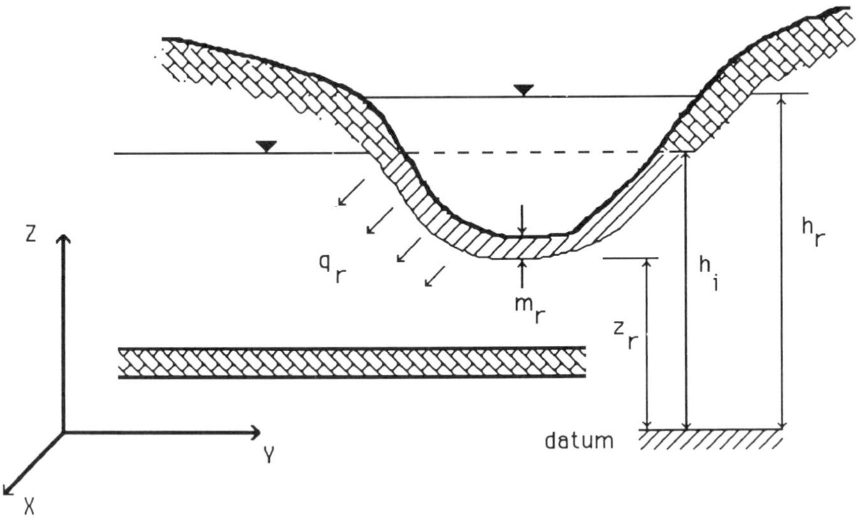

Figure 7. Definition sketch for a Cauchy-type (head-dependent) boundary condition.

In typical regional modeling studies, the head-dependent boundary condition defined above is not used as frequently as the other two boundary conditions. It is included in the computer code discussed in this book, with the following clarifications. First, for $h_i < z_r$, a constant flux condition is specified; thus, the conditions given in 2 and 3 above are equivalent. For such cases, the input procedures described for Neumann boundary conditions should be used at all times. Second, for $h_i > z_r$, the equation given above may be written as:

$$q_r = (K_r \, w_r \, h_r \,/\, m_r) - (K_r \, w_r \,/\, m_r) \, h_i \qquad (2.22)$$

which has a constant part (the first term on the right-hand side of the equality sign) and a head-dependent part (the second term on the right-hand side). For data input considerations, the first term should always be treated as a typical Neumann boundary condition. To complete the boundary condition input as given in

Equation 2.22, the constant in front of the piezometric head variable, h_i, in the second term should be used as the input boundary value for the boundary segments for which a head-dependent boundary condition is specified. This implies that the treatment of this type of boundary condition will require input data using both the Neumann boundary condition format and the head-dependent boundary condition format. As is the case with the previous two boundary conditions, the Cauchy-type boundary condition may be specified as a time-dependent condition in the computer code developed.

INTERPRETATION FOR CASE STUDIES

The model presented in this book may be used to analyze simple, fully confined regional aquifers with homogeneous or heterogeneous field parameters. The aquifer boundaries may be treated as constant piezometric head boundaries and/or constant flux or Cauchy-type boundaries in segments. The fully confined aquifer may be tapped by several wells, which may be treated as pumping or recharge wells. This example is a typical, and the most common, applications study and may be analyzed using the model described in this book. A definition sketch of such an aquifer is shown in Figure 8.

In addition to this simplified applications study, the mathematical model and the computer code in this book may be used in the analysis of more complex problems in ground water flow. With boundary conditions and well pattern similar to the one-layer aquifer described above, a leaky aquifer system may be treated up to a maximum of five main aquifer layers. A one-layer unconfined aquifer problem with infiltration, evaporation, or both may be analyzed, which may or may not include well networks. Alternatively, this unconfined aquifer may be treated as the top layer of a four-layer multilayer aquifer system in which the piezometric head distributions in the main aquifer layers are functions of the piezometric head conditions of adjacent aquifer layers. This coupling effect is introduced to the mathemati-

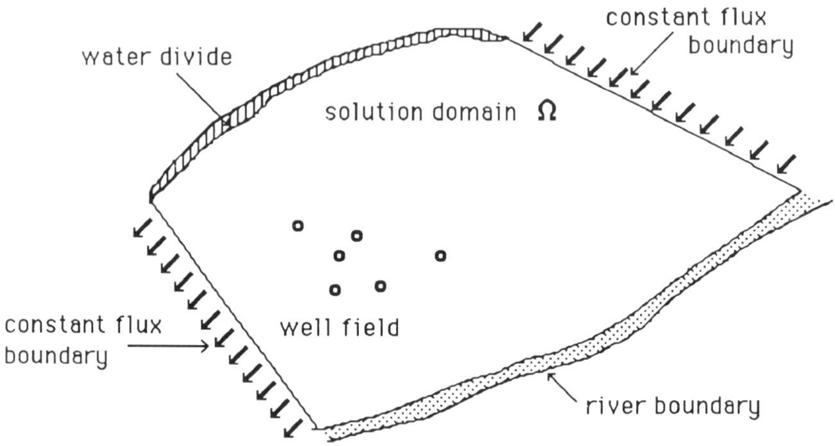

Figure 8. Definition sketch for a typical regional aquifer.

cal–numerical model through analytical solution of leakage terms.

Given this overall picture and not exceeding the dimension limits of the computer code described in Chapter 4, the user may decide upon the level of complexity of the problem that needs to be analyzed. For most applications, the data input requirements of the code are very simple but may increase proportionally to the complexity of the problem. As mentioned earlier, finite-element procedures are used to solve the governing partial differential equations discussed in this chapter. In certain applications, input data preparation for this method can be very time-consuming. To minimize the input data preparation time and effort, several internal data generation routines are included in the computer code. Compared with other finite-element data generation routines that are available in the literature and commercially, the routines in this book may be considered elementary. Although the data generation pattern discussed in Chapter 4 is simple, useful, and general enough to develop the necessary data for most complex applications, commercial data generation software compatible with the elements used in the numerical method discussed in Chapter 3 can also be used in input data generation for the code.

The numerical procedure used in the solution of the preceding partial differential equations is iterative. For an unconfined aquifer problem, iterative solutions are necessary due to the nonlinearity of the governing partial differential equations; however, in this book iteration is also employed in the solution of linear equations defined for confined or semiconfined aquifer systems.

These iterative procedures are used in the solution of global matrix equations. Considering the dimensions of the matrices generated in a typical multilayer aquifer study, an iterative solution is necessary if one wants to analyze a multilayer aquifer problem with a reasonable number of nodes and elements in each layer in a microcomputer environment (640K). Iterations are performed at the matrix solution level. Error introduced to the solution may be controlled through error criteria, which need to be supplied as input data.

The iterative process does *not* imply a layer-by-layer solution in which the continuity of leakage between layers is adjusted iteratively, as was done in some earlier applications. In this book, matrix equations are solved at once for all layers for the primary unknowns, h_i. Further details of the numerical aspects of this computation process and implications for data preparation are given in the following chapters.

3

Numerical Solution

The purpose of this chapter is to present the numerical formulation developed for the solution of the governing partial differential equations discussed in Chapter 2. Attention shall be restricted to unsteady ground water flow problems, since the formulation of steady-state analysis in multilayer aquifers was discussed earlier (Aral 1990). As in the book on steady flow, the numerical solution will be obtained by the finite-element method because of this method's advantages both in the formulation and the programming phases of a typical applications study. The literature available in the ground water modeling field demonstrates that the finite-element method lends itself quite favorably to aquifer analysis, enabling easy idealization of irregular aquifer boundaries, easy representation of heterogeneous aquifer properties, and simpler and accurate solution of nonlinear problems. The finite-element method has been applied to a wide variety of complex aquifer problems (Pinder and Gray 1977; Huyakorn and Pinder 1983).

In what follows, the basic steps involved in finite-element Galerkin formulation of unsteady flow in multilayer aquifers will be discussed. Further details of the theory of finite-element analysis may also be found in the above references.

THE FINITE-ELEMENT CONCEPT

Historically, the finite-element method started as a numerical tool in solid mechanics in the mid-1950s. First applications may be found in the analysis of structural systems as an alternative to classical mathematical techniques in dealing with complex geometrical configurations. The extension of the method to applications in continuum problems was made possible when it was realized that the method was strongly related to energy principles, which may be associated with variational techniques. With this extension, the application of finite-element analysis to disciplines other than structural mechanics was made possible. In fact, the technique could be used to approximate any mathematical or physical problem whose solution could be expressed as a stationary value of a given functional.

Although the variational approach has several advantages, it also has major limitations. In some applications, it may be extremely difficult to determine the variational formulation associated with a partial differential equation whose solution is required. In other instances, such as unsteady-flow problems or convection problems, although such variational forms may be defined in a restricted form, the numerical solution of the resulting integral equations may prove to be cumbersome and difficult. In finite-element formulations, this disadvantage in formulation was overcome by the use of the method of weighed residuals, specifically, the Galerkin method.

A typical analysis starts with the formulation of governing partial differential equations, as presented in Chapter 2.

The next step is the idealization of the solution domain by a finite number of distinct nonoverlapping regions, called *finite elements,* over which the unknown variables are to be interpolated. In this idealization, all of the properties of the original system, such as transmissivity, permeability, specific storage, etc., will be retained in the individual elements (Figure 9). This in turn makes the treatment of variable field properties possible, which in itself is one of the principal advantages of the finite-element method. Through this interpolation, the unknowns, i.e., the dependent variables of the mathematical model, are associated with the "nodes" of each element. Once a finite-element "type" is generated, a decision has been made on an approximation for the distribution of the unknown variable

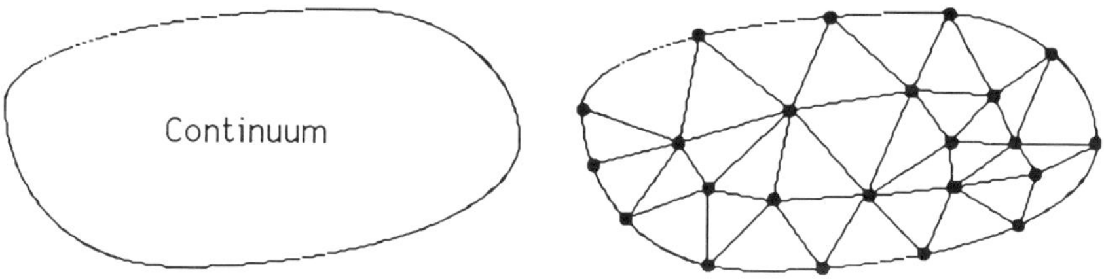

Figure 9. Idealization of a continuum by triangular elements.

over the element. More often than not, a Lagrangian type of approximation is employed in this process.

The next step is the derivation of element equations. There are several alternatives in this stage. The two most common techniques used are based on variational principles (energy method) or residual methods (method of weighed residuals). The more commonly adopted method, i.e., the method of weighed residuals, will be used to derive the finite-element matrix equations for the unsteady multilayer aquifer analysis in this text. Once the element equations have been evaluated, the resulting matrices are assembled by joining all elements at their respective junctions to form the complete analytical model of the continuum, applying, in the process, the requirement of point equilibrium and compatibility.

Matrix methods provide the most practical means of organizing the computations and assembly. It is a characteristic of the finite-element method, whether used in a structural sense or employed to analyze other phenomena, that the essential properties of an element will be of the form encountered in structural analysis.

Hence, after the formulation of element matrix equations, the general procedures of assembly and solution will follow a pattern based on structural analogy. High-speed computers are an essential part of the overall solution process. The assembly process yields a set of simultaneous algebraic equations that may be solved in a variety of ways. At the end of the matrix solution stage, one obtains an approximate solution to the mathematical problem described in the first step. This solution will be a discrete solution, i.e., the solution will be based on values at discrete points. What remains now is the evaluation of the secondary quantities such as velocities and pressures, which depend on the solution obtained.

The final step is the interpretation of the results obtained within the error bounds of the numerical algorithm.

The finite-element concept is described intuitively above. To summarize, the first approximation applied to the continuum is of a physical nature: a modified system is substituted for an actual continuum. There is no need for approximations in the differential equations of the substitute system. Approximations are introduced, however, in interpolating the unknown variable in an element. Thus, the steps involved in the analysis, as treated in this text, may be given as:

1. Formulation of the differential equation.
2. Division of the continuum by imaginary lines or surfaces into a number of elements.
3. Formulation of the integral equations within an element, using method of weighed residuals (or variational principles).
4. Formulation of the finite-element matrix equations within an element.
5. Assembly of element matrix equations for all elements.
6. Substitution of boundary conditions into the resultant matrix equations.
7. Solution of the resultant system of simultaneous ordinary differential equations in time.

These steps will be used later in expanded form in the numerical solution of the partial differential equations discussed in Chapter 2. A detailed explanation of the application of this method to continuum problems and an extensive historical review can be found in Zienkiewicz (1977). Application of the finite-element method to ground water modeling can be found in Pinder and Gray (1977) and in Huyakorn and Pinder (1983).

FINITE-ELEMENT FORMULATION

In developing a numerical solution for the mathematical model given in the previous chapter, a finite-element Galerkin process was adopted. Consider the governing equations for a multilayer aquifer system as defined in Equation 2.4 or 2.5. In operator form, these equations may be given as:

$$L(h_i) - f = 0 \qquad i = 1, 2, 3, ..., J \qquad (3.1)$$

where L represent the differential operator defined in these equations in a domain Ω_i enclosed by the boundary B_i (Figure 6); f may represent the source, sink, or leakage terms of the same equations; and J is the number of layers. According to finite-element Galerkin process, the domain Ω_i of each layer is discretized into a number of subdomains identified as elements, A_i^e. The next step is to approximate the unknown function, in this case the piezometric head distribution in each layer, h_i, over a subdomain, with a trial function of the form:

$$h_i \approx \bar{h}_i(x,y,t) = N_m(x,y)\, h_{i,m}(t) \qquad m = j, k, l \qquad (3.2)$$

where N_m are linearly independent basis functions and $h_{i,m}$ are the nodal values of the piezometric head at the nodes of the idealized domain, which is the ith layer in multilayer aquifer applications. As can be seen in Equation 3.2, space and time variations are separated in this interpolation. The basis functions are functions of space coordinates, and the unknown nodal values are considered to be a function of time. For a three-nodal triangular element, as used in this computer code, the basis functions may take the form:

$$N_m = (1/2A)\,(a_m + b_m x + c_m y) \qquad m = j, k, l \qquad (3.3)$$

in which A is the area of the triangular element (subdomain) and the interpolation polynomial is linear (Figure 10). The coefficients a_m, b_m, and c_m for m = j, k, l, are defined as follows:

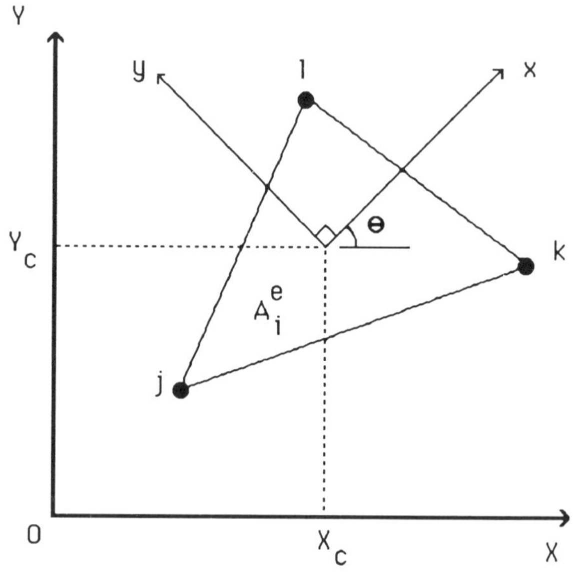

Figure 10. A three-nodal triangular element.

$$\left.\begin{aligned} a_j &= x_k\, y_l - x_l\, y_k \;;\; b_j = y_k - y_l \;;\; c_j = x_k - x_l \\ a_k &= x_l\, y_j - x_j\, y_l \;;\; b_k = y_l - y_j \;;\; c_k = x_l - x_j \\ a_l &= x_j\, y_k - x_k\, y_j \;;\; b_l = y_j - y_k \;;\; c_l = x_j - x_k \end{aligned}\right\} \quad (3.4)$$

and

$$A = (1/2)\, \mathrm{ABS}(a_j + a_k + a_l) \quad (3.5)$$

where x_m,y_m (m = j, k, l) represent the local coordinates of the triangular element. In developing element equations, in order to simplify the resultant equations, the origin of the local coordinate axis is placed at the centroid of the element and the principal axes are inclined in the direction of local anisotropy (Figure 10). This way tensor representation of aquifer permeabilities or transmissivities will be reduced and Darcy's law may be used in the form given in Equation 2.1.

Because the trial function $\bar{h}_i$ is only an approximation, Equation 3.1 will not be satisfied exactly. Substitution of Equation 3.2 into 3.1 thus results in an error or residual.

$$R_i = L(\bar{h}_i) - f \qquad i = 1, 2, 3, \ldots, J \tag{3.6}$$

The Galerkin method seeks to determine the unknown parameters $h_{i,m}$ in such a way that the error is minimal in the following sense:

$$\int_{\Omega_i} W_m\,(L(\bar{h}_i) - f)\,d\Omega_i = 0 \qquad \text{for} \begin{cases} m = 1, 2, 3, \ldots, n \\ i = 1, 2, 3, \ldots, J \end{cases} \tag{3.7}$$

in which W_m are the weighing functions that are chosen as the basis functions, as defined earlier, in the case of Galerkin formulation. In discrete form Equation 3.7 may be written as:

$$\sum_{e=1}^{n_e} \int_{A_i^e} N_m[L(\bar{h}_i) - f]\,dA_i^e = 0 \qquad \text{for} \begin{cases} m = 1, 2, 3 \\ i = 1, 2, 3, \ldots, J \end{cases} \tag{3.8}$$

where summation is implied over all elements. Substituting Equations 2.4 or 2.5 and 3.2 into Equation 3.8 and reducing the order of second-order differential terms through the use of Greens theorem, one can numerically integrate the resulting equations over an element, which yields the matrix equations given below for the ith aquifer layer:

$$[S_i^e]\,\{h_i^e\} + [M_i^e]\,\{dh_i/dt\} = \{F_i^e\} \tag{3.9}$$

where the coefficients of $[S_i^e]$, $[M_i^e]$, and $\{F_i^e\}$ can be given as:

$$(S_i^e)_{j,k} = \int_{A_i^e} [T_{xi}^e \frac{\partial N_j}{\partial x} \frac{\partial N_k}{\partial x} + T_{yi}^e \frac{\partial N_j}{\partial y} \frac{\partial N_k}{\partial y}] \, dA_i^e + \int_{A_i^e} N_j \, \bar{q}_i(h_i) \, dA_i^e \tag{3.10}$$

$$(M_i^e)_{j,k} = \int_{A_i^e} S_{si}^e \, N_j \, N_k \, dA_i^e \tag{3.11}$$

$$(F_i^e)_j = \sum_{k=1}^{w} \int_{A_i^e} N_j \, Q_{i,k}(x_k, y_k) \, dA_i^e + \int_{B_{2i}^e} N_j \, [\mathbf{n} \cdot \mathbf{T}_i^e \, \nabla h_i^e] \, dB_{2i}^e - \int_{A_i^e} N_j \, \bar{q}_i(h_{i-1}, h_{i+1}) \, dA_i^e \tag{3.12}$$

which are standard finite-element Galerkin matrix coefficients, assuming that the field properties T_{xi}, T_{yi} and S_{si} are constant in an element. For the case of an unconfined aquifer, Equation 3.10 takes the form:

$$(S_i^e)_{j,k} = \int_{A_i^e} [K_{xi}^e h_{i,ave}^e \frac{\partial N_j}{\partial x} \frac{\partial N_k}{\partial x} + K_{yi}^e h_{i,ave}^e \frac{\partial N_j}{\partial y} \frac{\partial N_k}{\partial y}] \, dA_i^e + \int_{A_i^e} N_j \, \bar{q}_i(h_i) \, dA_i^e \tag{3.13}$$

where $h_{i,ave}^e$ is the average of the nodal values of the piezometric head in each element at each iteration step. The iteration process used in this solution is explained in more detail in the following sections. Also, for the unconfined aquifer case, Equation 3.12 will take the form shown below.

$$(F_i^e)_j = \sum_{k=1}^{w} \int_{A_i^e} N_j\, Q_{i,k}(x_k,y_k)\, dA_i^e + \int_{B_{2i}^e} N_j\, [\mathbf{n} \cdot K_i \quad h_i]\, dB_{2i}^e$$

$$+ \int_{A_i^e} N_j\, \bar{q}_i(h_{i-1},h_{i+1})\, dA_i^e + \int_{A_i^e} N_j\, I_i^e\, dA_i^e \qquad (3.14)$$

In Equations 3.10 and 3.13, $\bar{q}_i(h_i)$ represent those terms of Equation 2.15 which are a function of h_i. One should realize that these terms contain h_i and $\bar{h}_i$, which represent the values of the piezometric head at two different time levels. Thus, in the generation of matrix equations, these terms should be properly placed into the $[S_i(t+\Delta t)]$ and $[S_i(t)]$ matrices, which are described in the following section. Similarly, in Equations 3.12 and 3.14, $\bar{q}_i(h_{i-1}, h_{i+1})$ represent those terms of Equation 2.15 that are a function of these variables. Since these terms are lumped into the proper time-dependent load vectors, $\{F_i(t+\Delta t)\}$ and $\{F_i(t)\}$, an iterative solution is now necessary in which the solution should be started with initially assumed values of $\bar{h}_{i-1}$, $\bar{h}_i$, and $\bar{h}_{i+1}$. Although there are other possible ways of treating the leakage terms, the formulation given above would lead to the most efficient and effective means of handling multilayer aquifer formulation in limited main core storage environments such as microcomputers (640K). In the overall solution, these iterations will be combined with the matrix iterative procedures, details of which are described in the following sections. The computation of Equations 3.10–3.14 is done assuming that K_{xi}^e, K_{yi}^e, $h_{i,ave}^e$, I_i^e are constant over an element.

The global assembly of these matrix equations for the ith layer yields:

$$[S_i]\,\{h_i\} + [M_i]\,\{dh_i/dt\} = \{F_i\} \qquad i = 1, 2, 3, ..., J \qquad (3.15)$$

These coupled matrix equations may now be combined in a single matrix system for J-layers as shown below.

$$[D]\,\{h\} + [P]\,\{dh/dt\} = \{G\} \qquad (3.16)$$

where

$$[D] = \begin{bmatrix} [S_1] & & & & & & & \\ & [S_2] & & & & & 0 & \\ & & \cdot & & & & & \\ & & & \cdot & & & & \\ & & & & \cdot & & & \\ & & & & & [S_i] & & \\ & & & & & & \cdot & \\ & & & & & & & \cdot \\ & & & & & & & & \cdot \\ & 0 & & & & & & & & [S_{J-1}] \\ & & & & & & & & & & [S_J] \end{bmatrix} \tag{3.17}$$

and

$$[P] = \begin{bmatrix} [M_1] & & & & & & & \\ & [M_2] & & & & & 0 & \\ & & \cdot & & & & & \\ & & & \cdot & & & & \\ & & & & \cdot & & & \\ & & & & & [M_i] & & \\ & & & & & & \cdot & \\ & & & & & & & \cdot \\ & & & & & & & & \cdot \\ & 0 & & & & & & & & [M_{J-1}] \\ & & & & & & & & & & [M_J] \end{bmatrix} \tag{3.18}$$

and

$$\{h\} = \left\{ \begin{matrix} \{h_1\} \\ \{h_2\} \\ \vdots \\ \{h_i\} \\ \vdots \\ \{h_J\} \end{matrix} \right\}, \quad \{dh/dt\} = \left\{ \begin{matrix} \{dh_1/dt\} \\ \{dh_2/dt\} \\ \vdots \\ \{dh_i/dt\} \\ \vdots \\ \{dh_J/dt\} \end{matrix} \right\} \tag{3.19}$$

and

$$\{G\} = \left\{ \begin{matrix} \{F_1\} \\ \{F_2\} \\ \vdots \\ \{F_i\} \\ \vdots \\ \{F_J\} \end{matrix} \right\} \tag{3.20}$$

At this point Neumann and head-dependent boundary conditions and leakage terms are already introduced to the matrix equations 3.12 and 3.14. After insertion of the Dirichlet boundary conditions for each layer, Equation 3.16 may now be solved iteratively for the unknown nodal values of the piezometric head in each aquifer layer at each time step. A finite-difference algorithm will be used for the integration of resultant simultaneous ordinary differential equations (3.16). In case of unconfined aquifer problems, the matrix [D] contains unknown nodal values of the piezometric heads, yielding a nonlinear simultaneous ordinary differential equation system (see Equation 3.13). Such a matrix equation will require an iterative solution also. However, the matrix solution process adopted here is already iterative, as discussed above for leakage terms. Thus, the iterative solution required for nonlinear equations will be combined with the iterative solution of matrix equations as described in the following sections. This solution process yields an efficient algorithm for the analysis of unsteady-state problems for a multilayer aquifer system in a microcomputer environment.

STEP-BY-STEP SOLUTION OF ORDINARY DIFFERENTIAL EQUATIONS IN TIME

The integration of Equation 3.16 is obtained in time dimension with a step-by-step technique suggested by Wilson and Clough (1962). In this algorithm the variation of $\{dh/dt\}$ is assumed to be linear in each time interval Δt. Thus, using Taylor series expansions for the primary unknown h_i, and ignoring the second and higher order terms in this expansion, the following equation can be derived, which reflects the value of $\{h\}$ at time $(t+\Delta t)$ in terms of the values of $\{h\}$ at time (t).

$$\{h_{t+\Delta t}\} = \{h_t\} + (\Delta t/2)\,[\,\{dh/dt_{t+\Delta t}\} + \{dh/dt_t\}\,] \quad (3.21)$$

Substituting Equation 3.16 into 3.21, the following recursive equation may be obtained which can be solved for the values of $\{h_{t+\Delta t}\}$ if the conditions at the beginning of the time step are known.

$$\left(\frac{2}{\Delta t}[P] + [D_{t+\Delta t}]\right)\{h_{t+\Delta t}\} = \left(\frac{2}{\Delta t}[P] - [D_t]\right)\{h_t\} + \{F_{t+\Delta t}\} + \{F_t\} \quad (3.22)$$

Since both the matrix [D] and the vector $\{F\}$ contain coefficients which are a function of unknown nodal values of the piezometric head variable, the step-by-step solution described by the above equation will be solved iteratively at each time step as described below. To emphasize a point made earlier: in Equation 3.22, the matrix [D] and the vector $\{F\}$ are evaluated at two different time levels. This implies that the proper time level leakage terms should be assembled into this matrix and vector system for a proper computational sequence, as described in the previous section.

In the solution of time-dependent problems, the proper choice of the time step (Δt) is extremely important. Especially in the solution on nonlinear problems, this choice may become very critical in achieving proper and fast convergence of iterations performed at each time step. To increase accuracy, small time steps are always recommended at the cost of increased computation time. In either case, a sensitivity analysis should always be performed in a typical study. This sensitivity analysis can be made on Δt and when comparing the two consecutive numerical results obtained. Significant differences between the two numerical solutions will imply improper convergence. In such cases a smaller (Δt) must be utilized to improve the accuracy of numerical results.

COMPUTATION OF THE ITERATIVE SOLUTION

The matrix equation generated above (3.22) can be solved directly for the piezometric head values in each aquifer layer for linear problems if the leakage terms were not lumped into the load vector. However, if one is working in a limited memory environment, such as 640K microcomputers, this task quickly becomes impossible, considering the dimensions of the matrix [P] and [D] for multilayer aquifer applications with at least 600 nodes, if a one-layer system is analyzed. In this computer code, this limit is considered to be reasonable for typical applications. The dimensions of this matrix can be reduced considerably if one adopts an iterative solution. Equation 3.16 may be rewritten, storing the matrices [P] and [D] in rectangular form, as described in the steady-state solution (Aral 1990). This new system can be given as:

$$[D_{new}]\,\{h\} + [P_{new}]\,\{dh/dt\} = \{G\} \tag{3.23}$$

where

$$[D_{new}] = \begin{bmatrix} [S_{1,new}] \\ [S_{2,new}] \\ \cdot \\ \cdot \\ \cdot \\ [S_{i,new}] \\ \cdot \\ \cdot \\ \cdot \\ [S_{J-1,new}] \\ [S_{J,new}] \end{bmatrix} \qquad (3.24)$$

and

$$[P_{new}] = \begin{bmatrix} [M_{1,new}] \\ [M_{2,new}] \\ \cdot \\ \cdot \\ \cdot \\ [M_{i,new}] \\ \cdot \\ \cdot \\ \cdot \\ [M_{J-1,new}] \\ [M_{J,new}] \end{bmatrix} \qquad (3.25)$$

with the vectors $\{h\}$, $\{dh/dt\}$, and $\{G\}$ are in the form given in Equations 3.19 and 3.20. To further reduce storage requirements, the matrices $[M_{i,new}]$ and $[S_{i,new}]$ are stored in a rectangular form, where only the nonzero upper diagonal band coefficients are retained for each layer. These are the new forms of the matrices $[M_i]$ and $[S_i]$, respectively. This can be done since the $[M_i]$ and $[S_i]$ matrices are symmetric matrices. In a typical solution process, only the coefficients in the upper band of these matrices will be required. The dimensions of the matrix $[P_{new}]$ and $[D_{new}]$ are now (no. of layers $\times$ no. of nodes in each layer $\times$ upper band width of the matrix $[M_i]$ and $[S_i]$), which shows considerable reduction in matrix size when compared to the size of the original matrices. Upper band width is defined as the maximum difference between the nodal numbers of each element plus 1. A more detailed discussion of this subject and operations required with matrices stored in this form is found in Cook (1974) and Carnahan et al. (1969). Now Equation 3.22 can only be solved using specialized simultaneous algebraic equation solvers that take into consideration the distorted positions of the coefficients of these matrices.

Given an initial piezometric head distribution and the assumed piezometric head distribution at time $(t+\Delta t)$, in each aquifer layer, the solution of Equation 3.23, using the procedure in Equation 3.22, may be given as:

$$\left(\frac{2}{\Delta t}[P_{new}] + [D_{new,t+\Delta t}]\right)\{h_{t+\Delta t}\} = \left(\frac{2}{\Delta t}[P_{new}] - [D_{new,t}]\right)\{h_t\} + \{G_{t+\Delta t}\} + \{G_t\} \quad (3.26)$$

This solution will yield the first iteration of the piezometric head values in all layers at the end the time interval Δt. Given this first iteration, the vector $\{G_{t+\Delta t}\}$ can now be updated. Repeating the solution of Equation 3.26 with the new right-hand side vector and the $[D_{t+\Delta t}]$ matrix completes the first cycle of the iterative process within the time interval Δt. The iterative process continues until a predetermined convergence limit is achieved. Convergence of iterations is measured using a euclidean norm given as:

$$\text{Error} = \left[\frac{\sum_{i=1}^{J}\sum_{j=1}^{n}\left(h_{ij}^{l+1} - h_{ij}^{l}\right)^2}{\sum_{i=1}^{J}\sum_{j=1}^{n}\left(h_{ij}^{l}\right)^2}\right]^{1/2} \quad (3.27)$$

in which the superscripts denote the iteration sequence, J is the number of layers, and n is the number of nodes in each layer. If the computed value of the error is smaller than a user-selected error limit, iterations will be stopped; otherwise the program continues with the iterative solution until a user-selected maximum iteration value is reached. If the maximum iteration value is reached before a true convergence is obtained, the program will be stopped and the results will be printed out with the value of the number of iterations performed and the error value in the last iteration. At this point, the user may decide to increase the maximum number of iterations specified in the input data (see Chapter 4) and reexecute the program to see whether the numerical results are changing appreciably with an increased number of iterations.

In the case of an unconfined aquifer problem, the matrix $[D_{new}]$ contains the unknown nodal values of the piezometric head in each element in each layer expressed by the variable h_{ave} (Equation 3.13), yielding a nonlinear simultaneous algebraic equation system. These nonlinear equations are solved using the Picard method, as is performed for the leakage terms. The general procedure for Picard iterative solution can be described as follows: Consider a general matrix equation similar to Equation 3.16, where the coefficients of [D] and $\{G\}$ are a function of $\{h\}$. In Picard solution, we start the iteration by assuming an initial solution $\{h\}^r$, where the superscript indicates the iteration sequence. Using this estimate the matrix [D] and the vector $\{G\}$ can be evaluated, which yield a set of linear equations when Equation 3.22 is considered. These equations can be solved for the next set of values of $\{h\}^{r+1}$. This provides a practical means for obtaining successive solutions of $\{h\}$. In this computation sequence, the matrix [D] and the vector $\{G\}$ are updated at each iteration cycle, and the iteration continues until satisfactory convergence is achieved. A typical check for convergence can again be done using the euclidean norm definition given in Equation 3.27. Further details of this computation process can be found in Carnahan et al. (1969) and Varga (1962). In this computer code, the Picard iteration required for the solution of nonlinear equations is combined with the iterations performed to solve the matrix system as described before. The general flow of operations used in ULAM.EXE code are summarized in Figure 11.

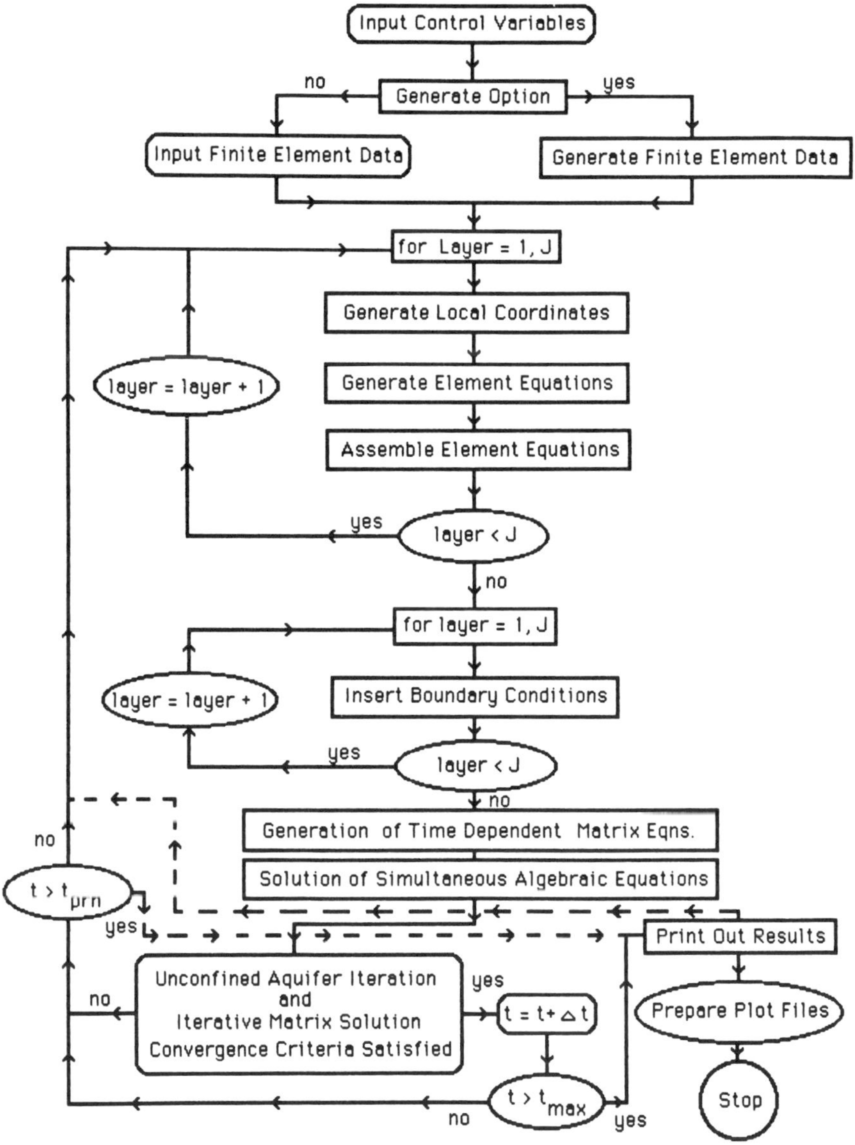

Figure 11. Flow of operations for the code ULAM.EXE.

4

User's Manual

The computer program presented in this book was written in Microsoft FORTRAN. The computer code ULAM.EXE (unsteady layered aquifer model) was developed as a tool for the practicing hydrogeologist to study unsteady-state conditions in a multilayer aquifer system. In order to present it in a form compatible with a wide variety of IBM PC or fully compatible microcomputers, the simplest format – 640K memory capacity and a low-density disk drive (360K) – is assumed available.

Due to the limitations on the memory capacity of IBM PC/AT, PS/2, or fully compatible microcomputers, the maximum number of layers, nodes, and elements that can be specified in a typical application is limited. The multilayer system may consist of a maximum of five layers – semiconfined, confined, or unconfined. For 640K memory capacity IBM PC/AT, PS/2, or fully compatible microcomputers with 360K disk drives, the computer code ULAM.EXE has the following limits:

- Number of layers cannot exceed 5.
- For a one- to five-layer system, number of elements in each layer cannot exceed 550.
- For a one-layer system, number of nodes cannot exceed 600.
- For a two-layer system, number of nodes in each layer cannot exceed 300.
- For a three-layer system, number of nodes in each layer cannot exceed 200.
- For a four-layer system, number of nodes in each layer cannot exceed 150.

- For a five-layer system, number of nodes in each layer cannot exceed 120.
- Number of Dirichlet, Neumann, and head-dependent boundary conditions in each layer cannot exceed 80.
- Upper band width, as defined in Chapter 3, cannot exceed 20.
- Number of wells in an aquifer layer cannot exceed 30.

For a 640K memory environment with a *high-density disk drive* or an *8087 math coprocessor*, or both, these limits may be extended. For these larger systems, a written request for the required version of the code may be placed with the author.

Although every effort has been made to present an error-free code through continuous upgrading, testing, and revising, the computer program will not be entirely free of bugs. The author anticipates further upgrading of the code in the future. If the user finds problems in implementation, or sees room for improvement or expansion, he or she should contact the author in writing at the address given in the screen message. Upgrades, high-density disk drive versions, or 8087 math coprocessor versions of the code may be requested from the author at the same address.

DISKETTE INSTRUCTIONS

A compiled executable file, ULAM.EXE (unsteady layered aquifer model), has been included in the distribution disk along with several sample data files that may be used to verify the computer code. The user should copy the file ULAM.EXE onto a working disk or, preferably, a hard disk.

In developing ULAM.EXE, it was assumed that the same two-dimensional finite-element idealization would be used in all layers in the x-y plane of each aquifer. A typical main aquifer layer may be idealized with triangular elements that may vary in size and geometric shape in the plane of the aquifer, as shown in Figure 1.

Except for a few changes, the data input format for unsteady flow analysis is the same as that for steady flow analysis (Aral

1990). The additional data required for unsteady flow analysis reflect the transient characteristics of the problem. In most cases, with a few minor changes, data sets prepared for steady flow analysis can be used for unsteady flow analysis.

Data as data files can be entered into ULAM.EXE in two ways: with the data generation option or supplying data as a whole (data input file option). *Data generation* is recommended for large, complex multilayer aquifer domains that require the preparation of large data files. If this option is selected, ULAM.EXE uses a base data set supplied by the user to generate the x,y coordinates of the element nodes, the finite-element connectivity matrix, transmissivity values, the angle of anisotropy, the thickness and permeability of the confining layers, the storage coefficient and specific storage of aquifers and aquitards, the upper and lower piezometric heads enveloping the main multilayer aquifer system, and the surface recharge and base elevations of the unconfined aquifer (in the case of a top free-surface aquifer). The *data input file* option is recommended for domains idealized by a small number of nodes and elements. If this option is chosen, the above-mentioned data should be supplied externally and no generation will be required.

A data file input format rather than screen menu input format was adopted because data must not be prepared hastily in front of a computer screen, which may result in errors and misrepresentations. Data preparation in a typical multilayer aquifer study is the most critical part of the overall work.

Input data files required to implement ULAM.EXE can be generated and edited using an editor available for the microcomputer system, or such files may be downloaded from mainframe computers. These data files should be checked very carefully before executing the program to avoid erroneous results.

The output from ULAM.EXE will be stored in files named INPUT.ECO, which will contain the echo of the input data, and UFGHSM.OUT, which will contain the unsteady-state solution of the piezometric head distribution in each layer. Darcy velocities will also be computed at each printout time interval for each layer, if requested, and will appear in UFGHSM.OUT. The output files may be printed out or viewed on the screen. Before results are analyzed, the printout of the input data (INPUT.ECO) should be checked very carefully to identify possible errors in the input data.

If the "plot" option was selected by the user, the x,y coordinates of each node in a main aquifer layer will be printed out, along with the piezometric head value at the node, as an X,Y,Z data set in a disk file named UPLOT?.DAT, at user-selected time intervals, in which the question mark will identify the layer number. These files may be used later for contour or surface plotting using compatible, commercially available computer codes, as demonstrated in Chapter 5.

The file ULAM.EXE may be executed by typing *ulam* and pressing ENTER at the DOS prompt. The user will be prompted by a screen message which will request an input data file name for "reader unit 5." At this point, the input data file should be on the hard disk path from which ULAM.EXE was run, or if the code was run from a disk drive A:\ or B:\ from a working disk, the data file should be available on the same disk drive. If a working disk on drive A:\ or B:\ was used, the user may replace the program disk with a data disk before entering the input data file name. (To stop execution of the program at this point, i.e., before entering the input data file name, the user should press CTRL-BREAK.) The program disk containing the file ULAM.EXE is no longer needed. The data disk, which has replaced the program disk, will now be used for further input/output operations. The run-time input data file should be on this new disk. Since the output files will also be stored on this disk, it should have enough empty storage space. After the first exchange of disks and after typing the unit 5 data file name and pressing ENTER, the user cannot and should not change disks. This process is recommended for problems where a lengthy output file is expected, besides which it will keep the input/output files of different problems separate.

A hard disk may be necessary to store output files in some cases of lengthy output. For example, the output from Problem 3 will not fit on a 360K disk if all output files are requested (as specified in the data file supplied with the distribution disk). If a hard disk is not available, the user should selectively choose the output files required or operate the code from a high-density floppy disk.

The next sections give instructions for preparing input data files, first by the data generation option and then by the data input option. The user is reminded that the input data file name should be the unit 5 reader identifier.

The ULAM.EXE file may also be used to generate another data file, IDEAL.DAT. This data file contains the idealization data for the aquifer. In addition to ULAM.EXE file, the distribution disk contains a utility program called IDEAL.BAS (IDEAL.EXE – see Appendix D), which may be used to plot the idealization of the solution domain using the data file IDEAL.DAT. IDEAL.EXE code is the Q-BASIC compiled version of the IDEAL.BAS file. Execution of this file is possible only if Q-BASIC compiler is available. (Q-BASIC compiler is *not* supplied in the distribution disk.) Plotting of the solution domain is much faster with this version if the user has access to Q-BASIC compiler. Otherwise IDEAL.BAS code can always be compiled using a standard BASIC compiler, and the code can be executed in a way similar to that described in the steady-flow analysis text. How to generate and use these data files is described below and in Chapter 5.

DATA GENERATION FORMAT

If the user chooses to *generate* a major portion of the input data, the data generation format should be chosen. The instructions should be followed in the order given. Sample data lines are given in display. A complete printout of the sample data files, included in the distribution disk, is given in Appendix B. A list of run-time error messages that may be printed out by the computer code is given in Appendix A with brief explanations.

1. **Select identifiers**. The computer code requires a *title* statement and a *units* statement as identifiers. Any consistent

set of units may be used. Each statement should not exceed 80 characters on a line, including blank spaces. The statements should be given on separate lines.

This is a test DATA.
Units are in feet, days.

2. **Select options.** A typical set of values may be given on one line, with the data separated by commas:

2,16,0,0,40,56,30,0.0001,0,1,1,0,1,1,150,3

This data line states that the user is specifying two confined or semiconfined layers, 16 Dirichlet boundary conditions in *each layer,* no Neumann or head-dependent boundary conditions, 40 nodes in each layer, 56 elements in each layer, a maximum of 30 iterations, 0.0001 as the convergence error criterion, units feet (or meters) with no conversion required, the generation option, an unconfined aquifer layer (the total number of layers specified is now three), no generation of plot data files, generation of discharge per unit width and idealization data files, a maximum time limit of 150 real-time units (seconds, days, etc.), and no flow in the multilayer aquifer system initially.

To prepare this data line, specify the following in the order given below.

a. **Number of confined layers in the multilayer aquifer system.** First layer is the bottom layer.
b. **Dirichlet boundary condition counter.**
c. **Neumann boundary condition counter.**
d. **Head-dependent boundary condition counter.**

The Dirichlet, Neumann, and head-dependent boundary counters specified above operate in the following way: If a number greater than 0 is specified (e.g., 16 Dirichlet conditions as specified above), it is implied that each layer has the same number of boundary conditions.

If a value of 0 is specified for these control variables, it is implied that the problem defined has no such boundary conditions in any of the layers. However, if a −1 is specified for any of these boundary condition counters, the user is instructing the code that each layer has a *different* number of boundary conditions for the specific boundary condition considered. In this case, the user has to specify the number of Dirichlet, Neumann, or head-dependent boundary conditions in each layer later in the data file before the boundary values in each layer are read in (see example problem 3B, Chapter 5, steady-state model [Aral 1990]). In the data line given above, it is implied that each of the three main aquifer layers has 16 Dirichlet boundary conditions and no Neumann or head-dependent boundary conditions.

e. **Number of nodes in each layer**. Since the number of nodes in each aquifer layer is the same for all layers, this value needs to be entered only once.
f. **Number of elements in each layer**. Since the number of elements in each aquifer layer is the same for all layers, this value needs to entered only once.
g. **Upper limit of number of iterations to be performed**. Number of iterations performed is a function of the error criterion, which is specified next.
h. **The error criterion**. Convergence of the iterative solution is checked using a euclidean norm comparing the two consecutive piezometric head solutions. Selecting a smaller error criterion implies more accurate results. Such a solution will require more iterations, the maximum of which is limited by the above input data (see 2g).
i. **Mile-to-foot or kilometer-to-meter conversion option**. Type *1* if x,y coordinate data for nodes are to be converted from kilometers to meters. Type *0* if no conversion is desired. Type any other value if miles are to be converted to feet. The user may enter the node coordinates in miles or kilometers and then request a conversion to feet

or meters. This may prove convenient if large aquifers are idealized.

j. **Data generation option**. Type *1* if data will be generated. Type *0* when a prepared set of input data will be used.

k. **Unconfined aquifer layer option**. Type *1* if the multilayer aquifer has a free-surface aquifer as the top layer. Type *0* if the aquifer system has only semiconfined or confined aquifer layers.

l. **Plot data file preparation option**. Type *1* if plotting data files need to be prepared. Type *0* if such data files will not be needed. This data file may be used to plot surface or contour plots of piezometric heads in each layer using commercially available software, as demonstrated in Chapter 5. These data files are named UPLOT?.DAT, in which the question mark is the layer number. The integer variable used in this control is identified as IPLOT and will appear on a screen message when the program is executed.

m. **Discharge per unit width data file preparation option**. Type *1* if discharge per unit width data files need to be prepared. Type *0* if such data files are not required. If *1* is selected, discharge per unit width and Darcy velocities at the centroid of each element in each main layer and confining aquifer layers will be computed and printed in a data file named UFGHSM.OUT, along with the coordinates of the centroid of each element, the total discharge per unit width, and the angle of inclination of the total discharge. The integer variable used in this control is identified as IVEL, and it will appear on a screen message when the program is executed.

n. **Idealization data file preparation option**. Type *1* if idealization data files are to be prepared. Type *0* if such data files are not required. This data file is named IDEAL.DAT. The distribution disk accompanying this book contains a program written in BASIC (IDEAL.BAS), which may be used to plot the finite-element idealization chosen on the screen or on hard copy (use the PRINT SCREEN command). This program should be run with BASIC or BASICA compilers available in every IBM or fully compatible microcomputer. Using this utility program and the screen plot of the idealization, one may immediately check the idealization selected and alter it if not satisfactory. Examples of the use of this program are shown in Chapter 5. The integer variable used in this control is identified as IDSCR, and it will appear on a screen message when the program is executed.

o. **Maximum time limit.** This input specifies the maximum real-time analysis limit of the unsteady-state solution, at which time the computations will be stopped.

p. **Initial condition control variable.** Type *3* if there is no flow in the multilayer aquifer system and the piezometric head distributions in all aquifer layers are constant and equal to the Dirichlet boundary conditions defined at the boundaries of the aquifers. Otherwise, type *1*, which implies that there is steady flow in the multilayer aquifer system, mainly determined by boundary conditions and well discharges, which needs to be solved for *before* starting the unsteady-state solution.

3. **Select the number of reference nodes and reference element boundaries in the main idealization directions.** The format should be as follows:

15,7,4

This data line states that the idealization chosen has 15 reference nodes, seven reference subintervals in idealization direction 1, and four reference subintervals in idealization direction 2.

To prepare this data line, first prepare a sketch of the idealization (see Figure 12). It is assumed that the distance between the circled nodes in direction 1 is divided into equal subinterval element boundaries, which are function node numbers assigned to the circled nodes. These circled nodes

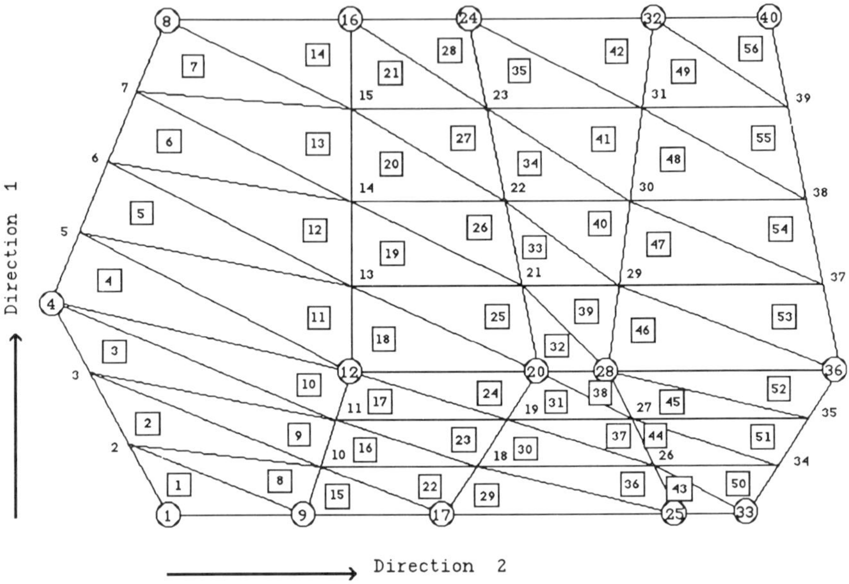

Figure 12. Idealization of a typical domain.

will be referred to as *reference nodes*. Thus, given the idealization shown in Figure 12, there are 15 reference nodes (node numbers 1, 4, 8, 9, 12, 16, 17, 20, 24, 25, 28, 32, 33, 36, 40). During data generation, the distance between nodes 1 and 4, for example, will be divided into three equal subintervals (4 - 1 = 3) and nodes 2 and 3 will be placed equal distances between reference nodes 1 and 4. Similar generation patterns will be repeated for all nodes in between the reference nodes in direction 1.

Since the number and position of reference nodes are totally up to the user, a complex boundary geometry or variable element size distribution within the aquifer domain, or both, may be selected without much difficulty, as demonstrated in Figure 12. This type of idealization pattern will yield a fixed number of subinterval (element boundaries) in main idealization directions 1 and 2, as shown in

Figure 12. These subintervals will be identified as *reference subintervals* in main idealization directions 1 and 2. In the idealization shown in Figure 12, there are seven reference subintervals in the idealization direction 1 and there are four reference subintervals in the idealization direction 2. Thus, *7* and *4* should also be entered, as shown in the above statement, for proper generation of the connectivity matrix.

4. **Generate the x,y coordinates of all the nodes in all aquifer layers and the finite-element connectivity matrix of all layers.** First specify x,y coordinates of the reference nodes. This data set should be supplied in increasing order and in the following format:

N,X(N),Y(N),node missed

where N is the reference node number, X(N) and Y(N) are the reference global node coordinates, and "node missed" is an integer variable with a value 1 or 0, which indicates whether generation is requested between this reference node and the next. If "node missed" is read in as 0, no generation will be performed between reference nodes N and N+1. Thus, reference node numbers N and N+1 should be two consecutive integers. However, if a value of 1 is specified, x,y coordinates of intermediate nodes will be generated between reference nodes N and N+1.

Next prepare the following 15-reference-node coordinate data set (for proper generation of the idealization shown in Figure 12):

```
1,0.0,0.0,1
4,-1.0,4.0,1
8,0.0,7.0,0
9,2.0,0.0,1
. . . . . . . .
40,8.0,6.0,0
```

With this data set, the computer code will generate the x,y coordinates of all the nodes in all aquifer layers and the finite-element connectivity matrix of all layers, which are basically the same. The data sets following this phase should be specified for each layer separately starting with the lowest aquifer layer, which is assumed to be the first layer.

5. **Generate the transmissivity values.** Generation is done by repeating a given data value for as many elements as desired. On one data line, specify the number of repetitions desired, the x-directional transmissivity, and the y-directional transmissivity. For example, given the idealization shown in Figure 12, the data line

55,60.0,60.0

repeats the x-transmissivity and y-transmissivity values of 60.0 units 55 times. Note that if the user selects 55 repetitions, then the first 56 elements will have the same value. This is the total number of elements used in the idealization shown in Figure 12. For example, in an idealization with 100 elements,

49,45.0,45.0
19,50.0,50.0
29,60.0,60.0

assigns a value of 45.0 units to the first 50 elements (elements 1–50), a value of 50.0 units for the next 20 elements (elements 51–70), and a value of 60.0 units to the next 30 elements (elements 71–100) within the same aquifer layer.

The element and node order numbering assumed in this generation pattern is shown in Figure 12. When the generation option is activated, the pattern generated will always follow the pattern shown in Figure 12. Generation of field parameters should proceed through all layers. If an application has three layers, as specified above, one should include all three aquifer transmissivities or permeabilities for unconfined aquifer layers. For example, for a three-layer aquifer system with 56 elements in each layer, as described above, the data set given below,

29,60.0,60.0
25,70.0,70.0
55,65.0,65.0
55,85.0,85.0

assigns an x and y directional transmissivity value of 60.0 units to the first 30 elements and a value of 70.0 units to the remaining 26 elements in the first layer. The middle layer has a uniform transmissivity value of 65.0 units, and the top layer has a uniform permeability value of 85.0 units. This simple way of generating heterogeneous aquifer properties may be used in all other input data, described below, which follow the same data input format. If the top layer is an unconfined aquifer, the values given for the top layer represent the permeabilities of the aquifer in respective directions in proper units.

6. **Generate the vertical permeability data of the confining aquifer layers.** This procedure is similar to the procedure described above. In applications where one does not have a free-surface aquifer, the number of confining layers should always exceed the number of main confined aquifer layers by one. For example,

55,1.5
55,1.5
55,1.5

assigns a vertical permeability value of 1.5 units for all three confining layers for all 56 elements in this sample data set.

7. **Enter the number of repetitions and the value of the thick-**

ness of the confining layers. This data set should be given for all confining layers. For example,

55,1000.0
55,1000.0
55,1000.0

This data set assigns a confining layer thickness of 1000 units for all 56 elements for all confining layers. The number of confining layers is always one more than the number of main layers, but in this case the top aquifer layer is specified to be an unconfined aquifer (input data line 3 above). Thus, the number of main aquifer layers and the confining aquifer layers should be the same in this application.

8. **Enter the number of repetitions and the value of the specific storage of the confining layers**. This data set should be given for all confining layers, starting with the bottom layer. For example,

55,0.00001
55,0.00001
55,0.00001

This data set assigns a specific storage of 0.00001 units for all 56 elements for all three confining layers.

9. **For each aquifer layer, enter the number of repetitions and the value of the storage coefficient in each element**. For example,

55,0.0005
55,0.0005
55,0.0005

implies that for all three aquifer layers, the storage coefficient is 0.0005 for all elements.

10. **For each main aquifer layer, enter the number of repetitions and the value of the angle of anisotropy in each element.** For example,

55,0.0
55,0.0
55,0.0

implies that for all three aquifer layers, the angle of anisotropy is 0.0.

11. **In the case of a free-surface aquifer, enter the number of repetitions and the recharge (or evaporation) value to be repeated for all elements.** For example,

55,0.003

implies a uniform recharge of 0.003 units over the aquifer. A negative value will imply evaporation. If the application does not have an unconfined aquifer layer, omit this data set and proceed with the input data set described in 13.

12. **In the case of a free-surface aquifer, give the elevation of the base of the aquifer layer. For example,**

39,78.

implies an elevation of 78.0 units for all the nodes of the free-surface aquifer. Notice that these data should be given for all *nodes* but not elements. This option allows the user to specify a variable base layer elevation for the unconfined aquifer, but considering the assumptions specified in Chapter 2, these elevation differences should not be large because uniform horizontal flow in all main layers is assumed throughout the analysis.

13. **Enter the number of repetitions and the upper and lower enveloping aquifer layer piezometric head values.** These are

the values of the piezometric heads at the *nodes* of an element and are not specified for elements. Thus, for the idealization shown in Figure 12, the following data set should be entered:

39,0.0
39,100.0

This data set assigns a piezometric head value of 0.0 units for all nodes on the lower enveloping layer and a piezometric head value of 100.0 units for all nodes on the upper enveloping layer.

14. **Specify the number of pumping and/or recharge wells.** On one data line, specify the number of wells in each layer. For example, for a three-aquifer-layer system, with two wells in the first layer and no wells in the remaining layers, the following data line should be prepared:

2,0,0

15. **For each of the confined layers and the unconfined layer, when present, supply the following well node and discharge data set.** Given the information above, two well nodes and two discharge values must be entered for the first layer. For example,

12,28
250.0,1250.0

implies that nodes 12 and 28 (in Figure 12) are assigned as well positions, and the corresponding discharge values for these wells are 250.0 units and 1250.0 units for the first layer. These data should be omitted if there are no wells specified.

16. **Enter the boundary condition values.** Enter Dirichlet first, Neumann second, and head-dependent third.

a. **Dirichlet**. Enter the node number, which specifies the boundary location, then the boundary condition value. For example,

1,100.,2,100.,3,100.,4,100.,5,100.,6,100.,7,100.,8,100.
33,20.,34,20.,35,20.,36,20.,37,20.,38,20.,39,20.,40,20.

This data set will assign a Dirichlet value of 100.0 units to the left-hand boundary of the idealization and a Dirichlet value of 20.0 units to the right-hand boundary (Figure 12).

b. **Neumann**. If specified on input line 3, declare the two boundary nodes involved for each element boundary, and the Neumann boundary condition. For example,

1,9,5.0
9,17,5.0
17,25,6.0

implies that there is a flux of 5.0 units between nodes 1 and 17 and a flux of 6.0 units between nodes 17 and 25.

By assigning Dirichlet, Neumann, or head-dependent boundary condition counters a value of −1 on data line 3, as described earlier, the user is instructing the computer code there is a different number of boundary conditions in each layer. In this case, the number of Dirichlet or Neumann boundary conditions should be entered before specifying the boundary values and nodes for each layer. For example,

10 (Dirichlet boundary condition counter for the first layer)
1,100.,2,100.,3,100., ...

or

3 (Neumann boundary condition counter for first layer)
1,9,5.0
. . . . .

c. **Head-dependent**. Follow the same procedure described for Neumann boundary conditions.

Items 15 and 16 should be repeated for each layer in sequence.

17. **Specify the time step used (Δt), the maximum real-time limit this time step will be utilized, and the printout interval within the maximum real-time interval**. For example,

0.01,50,100

implies that the time step specified is 0.01 time units for up to 50 time units the Δt given will be utilized, and the computed results will be printed out into an output file at every $100 \times \Delta t$ time intervals. Since the maximum real-time limit is specified as 50 in this data line, which is less than 150 as specified as the final time limit in the input data line 3, item 2, additional data are now necessary. After the completion of unsteady solution up to the time limit 50, the program will expect another input data set for the same variables. This input will be in the same format as the one described here. These data would not be necessary if one specifies a maximum time limit greater than the maximum defined in input line 3. This control is introduced in order to give the user an option to change the Δt value at different phases of the unsteady-state solution. Larger time intervals may be used in time integration of the governing equations as the solution approaches steady state without affecting stability and convergence of the numerical solution adversely. As expected, the control of Δt will save considerable computation time and printout space in most application studies. The next input of Δt, T_{max}, printout interval, data line (item 17), should be entered following the next data set, 18. For examples of typical data sets utilizing these controls, see Appendix B and Chapter 5.

18. **Enter transient-state data**. In a typical ground water flow

study, a transient state may be initiated by changing either (1) the pumping rates, (2) infiltration or evaporation, or (3) the boundary conditions of the problem as a function of time. These changes may now be introduced into the code as follows:

a. **Enter the time limit up to which these changes will be kept constant**. For example,

   75.0

   implies that, up to 75 (units of time), the conditions specified below will be kept constant. If the time interval specified is greater than the real-time solution time interval limit defined in input line 3, item 2, it is implied that these changes will be kept constant throughout the computation.

b. **Indicate whether there are changes in pumping rates control, infiltration and evaporation rates control, or boundary conditions control**. These control variables may be entered either as *0*, implying no change, or *1*, implying a change in the respective conditions. For example, the input data line

   1,1,1

   will imply a change in all three conditions for the period specified in 18a.

c. **If the first control in the above line is 1, enter new pumping rates and/or new pumping wells, as described in items 14 and 15**. For example, the input lines

   0,2,0
   13,22
   250.0,1250.0

   imply that, at the beginning of this new transient state, the two wells in the first layer specified earlier are shut down. Two new wells in aquifer 2 have been started at new nodes 13 and 22 for which the pumping rates are 250.0 and 1250.0 units. This data set should be supplied for all aquifer layers where a well is specified (see line one, 18c). This transient state will be kept constant up to the time limit 75 time units, as specified in 18a. This data set should be omitted if the first control in 18b is 0.

d. **If the second control in 18b is 1, enter new evaporation or infiltration values, as described in item 11.** This change can only be introduced if an unconfined aquifer is specified.

55,0.05

This data line implies that the earlier defined infiltration rate (0.003 units, item 11) is changed to 0.05 units for all elements. This data set should be omitted if the second control in 18b is 0.

e. **If the third control in 18b is 1, enter new boundary condition data.** Give Dirichlet boundary conditions first, then Neumann, then head-dependent, as defined in 16. If a (−1) is specified on data line 3 (items 2b, 2c, 2d) for these boundary conditions, then specify the number of boundary conditions first, as described in item 16b. For a transient boundary condition input, in which the number of boundary conditions in each layer are to be changed, then a (−1) should always be used on data line 3 (items 2b, 2c, 2d). This data set should be omitted if the third control in 18b is 0.

The data set described in 18 needs to be repeated as many times as necessary, depending on the time limits specified for each change and the maximum time limit defined in input line 3, item 2.

By following these procedures, the user creates a data file that enables ULAM.EXE to generate most of the required data sets. If the data generated yields run-time errors, the user should first check the data file. Appendix A gives a list of possible errors and error messages one might encounter. The generation option is flexible enough to develop data sets for very complex applications. The user should first practice with the data files supplied with the distribution disk to become more familiar with the patterns and sequences employed. Practice is also encouraged with the procedures described in the next section.

DATA INPUT FILE FORMAT

The data generation procedure described in this section is recommended for small domains idealized by a small number of nodes and elements. In this option, the computer code will not generate the x,y coordinates of the nodes or the connectivity matrix. The format of this section is the same as that of the previous section. Only differences in the steps will be given.

1. **Same as item 1.**

2. **Same as item 2, with one exception.** When entering the generation option (see 2j), the user should type *0*, implying that the x,y coordinates and connectivity matrix will be supplied as input data.

3. **Enter nodal coordinate data set and connectivity matrix data set generated in item 2 as input data set.** Enter the x,y coordinates of all nodes of the finite-element idealization in the format given below. These data should be supplied only for one layer. The computer program uses the same idealization in all layers. The data set should be sequential in increasing order, but in this case any idealization may be chosen, i.e., the idealization pattern used in the previous section need not be followed – any other idealization pattern may be chosen. However, the limits of the computer code, i.e., upper band width and number of node limits, should not be exceeded. For example,

   x(1),y(1)
   x(2),y(2)
   x(3),y(3)

....
x(N),y(N)

where N is the total number of nodes. This value is 40 for the idealization shown in Figure 12.

Next enter the connectivity matrix, which lists the node numbers of each element counterclockwise. Again, this data set should be prepared for one layer. The data set should be sequential in element numbering in increasing order, but as is the case with the node coordinate data input format described above, any idealization pattern may be chosen. This idealization pattern should yield input data within the limits of the computer code as defined previously, and the input data should be in the format given below:

1,9,2
2,10,3
3,11,4
......
39,40,32

for the idealization shown in Figure 12. A total of 56 data lines will be required for this particular idealization.

4. **Enter the x,y directional transmissivity values and the angle of anisotropy values.** For each layer, enter the x then the y directional transmissivity values in the following format.

$T_x(1)$, $T_x(2)$, $T_x(3)$, . . . , T_x(Number of Elements)
$T_y(1)$, $T_y(2)$, $T_y(3)$, . . . , T_y(Number of Elements)

Then enter the angle of anisotropy values.

ANG(1), ANG(2), ANG(3), . . . , ANG(Number of Elements)

Then enter the storage coefficients.

SC(1), SC(2), SC(3), . . . , SC(Number of Elements)

This data set should be repeated for all layers in the same format.

5. **Enter the vertical permeability of each confining layer followed by the thickness of each layer.**

K_z(1), K_z(2), K_z(3), . . . , K_z(Number of Elements)
D_z(1), D_z(2), D_z(3), . . . , D_z(Number of Elements)
S_s(1), S_s(2), S_s(3), . . . , S_s(Number of Elements)

This data set should be supplied for all confining layers in the same format.

6. **If the application has an unconfined aquifer, enter the surface recharge or evaporation data in the format shown below:**

I(1), I(2), I(3), . . . , I(Number of Elements)

This data set should be omitted if the multilayer aquifer system is a fully confined system.

7. **If the application has an unconfined aquifer, enter the base elevation of this layer in the format shown below:**

BS(1),BS(2),BS(3), . . . , BS(Number of Nodes)

This data set should be omitted if the multilayer aquifer system is a fully confined system.

8. **Enter the piezometric head values of the upper and lower enveloping layers.** These values should be given for each node in the following format:

LH(1), LH(2), LH(3), . . . , LH(Number of Nodes)
UH(1), UH(2), UH(3), . . . , UH(Number of Nodes)

9. **Same as item 14.**
10. **Same as item 15.**
11. **Same as item 16.**
12. **Same as item 17.**
13. **Same as item 18.**

OUTPUT ORGANIZATION AND PLOT FILES

The output from ULAM.EXE file will be stored in four output data files if all are requested by the user. These files are identified as INPUT.ECO, where an echo of the input data file will be stored; UFGHSM.OUT, where the computed unsteady results will be stored; UPLOT?.DAT, where piezometric head values are stored to be used later for plotting contour or surface plots; and IDEAL.DAT where idealization data will be stored to plot the idealization chosen on the screen.

The output file UFGHSM.OUT will mainly contain the printout of the unsteady-state piezometric head solution in each aquifer layer as well as the Darcy velocities times aquifer thickness at each requested time interval. The file is organized with proper headings and titles, and it is very easy to identify the individual output sequences. As a first step in the analysis, the user should check and control the INPUT.ECO output file very carefully and make sure that the input data is properly introduced to the computer code. If no other data file is requested, as explained in previous sections, these will be the only output data files prepared.

If requested, a third data file UPLOT?.DAT will be generated. This data file may be used for plotting the piezometric head distribution in each aquifer layer at respective times. The question mark will be replaced by the layer number in a regular output. This data file contains the x and y coordinates of all nodes in the aquifer and the value of the piezometric head at each node in an X,Y,Z data format. This format is the most common format required by commercially available plotting routines that may generate a contour plot or a surface plot of the output data. Some other output identification lines will be printed in this output data file that summarize time interval or other output file characteristics. Before using this file as an input data file for such software, these lines *should be erased.* This can be performed using a standard editor available on the microcomputer.

If requested, a fourth data file, the idealization data file (IDEAL.DAT), will be prepared. This file contains the x,y coordinates of all nodes and the connectivity matrix used in the finite-element idealization. IDEAL.DAT may be used directly as the input data file for the utility program IDEAL.BAS included on the distribution disk (IDEAL.EXE – see Appendix D) to plot the finite-element idealization of the aquifer. This plot will provide the user with a tool for immediate inspection of the idealization chosen. If the characteristics of this idealization – the distribution and size of elements chosen at specific points in the aquifer – are not satisfactory, the user may decide to change the idealization at this point without going further into the details of the analysis. A hard copy of this plot may also be obtained with the PRINT SCREEN command if a compatible printer is available. The program IDEAL.BAS is written in BASIC language following the outline given in Kinzelbach (1986), and thus it should be run through a BASIC or BASICA compiler available with every IBM or IBM-compatible personal computer.

This program is written for screen dimensions 640 by 480. If the user has a different graphics display format, these values should be changed in the program to match the screen sensitivity available. Statements containing these values can be seen on a printout of the program, or these changes can be made on the screen. In Chapter 5, the processes described in this chapter are used to analyze several demonstration problems.

5

Case Studies

In present-day geohydrologic analysis, numerical models are an essential part of most ground water studies because they provide an effective means of rapidly testing and evaluating a variety of scenarios for a given problem. To assure accuracy and efficiency in these studies, considerable research time and effort was invested toward the development of high quality numerical simulation procedures in earlier studies. At present, among other numerical procedures, the finite-element method seems to be the most popular computational technique for ground water applications.

The versatility of finite-element analysis in application to ground water studies becomes evident when one attempts to model irregular boundaries, nonhomogeneous and anisotropic aquifers, and nonlinear equations. Examples in this chapter were chosen to demonstrate how well the computer code ULAM.EXE handles these complexities and to help the reader become familiar with the input data preparation phase described in Chapter 4. The one-layer and multilayer problems discussed in this chapter were chosen to show that numerical results obtained with analytical or other numerical procedures published in the current literature are verifiable by this computer code. Other applications, which utilize earlier versions of this computer code, can be found in Aral (1985, 1986). The input data files used in these examples are on the distribution disks (without explanations) and in Appendix B (with explanations for each line of input data with cross-references to the procedures described in Chapter 4).

Application studies are grouped according to idealization and aquifer domain used. The different data input files in the various problems address different aspects of the input data file generation format described in detail in Chapter 4, but the example problems discussed in this chapter by no means exhaust the type of problem that may be analyzed by ULAM.EXE. Once the user becomes familiar with the characteristics of the overall code, applications other than the ones discussed in this chapter will become self-evident.

Numerical results obtained are shown either in graphic form (x,y, plots) or as topographic and/or surface plots. The x,y plots can be generated using a graphics (x,y) plotter that is commercially available. The piezometric head contour plots and surface plots are obtained by using the data files generated by the ULAM.EXE computer code and by the Golden Graphics Software program package, which is also commercially available. The plot of the idealization of the domain with triangular elements was obtained by the IDEAL.BAS (or IDEAL.EXE) computer code and the data file IDEAL.DAT. IDEAL.BAS is on the distribution disk and should be run using a BASIC or BASICA compiler, or it can be run directly, if IDEAL.EXE is used, by typing *ideal* and pressing ENTER at the DOS prompt if a Q-BASIC compiler is available (BRUN20.EXE file). A printout of the IDEAL.BAS computer code is given in Appendix D, which also includes the necessary information to adopt the program to various microcomputer environments.

PROBLEM 1: ONE-LAYER, UNCONFINED AQUIFER

In this problem, the movement of the water table caused by uniform recharge is investigated. The flow region is assumed to be saturated, and the unconfined aquifer is underlain by an impervious bottom. A definition sketch for the two-dimensional flow region in the plane of the aquifer is shown in Figure 13. The recharge is applied to a finite region ($0 \leq x \leq 23.8$ cm and $0 \leq y$

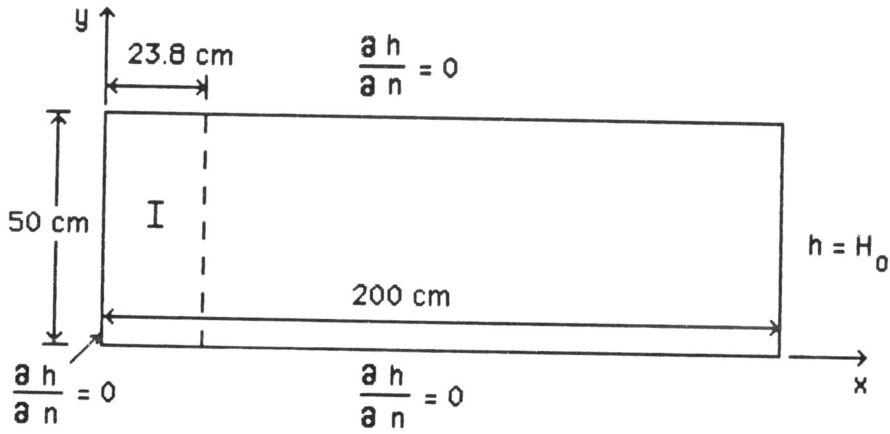

Figure 13. Definition sketch for Problem 1.

≤ 50 cm). The analytical solution used in this problem to verify the numerical results was developed by Marino (1967) for the analysis of growth and decay of a mound due to recharge. This solution depends on the experiments performed on a Hele-Shaw

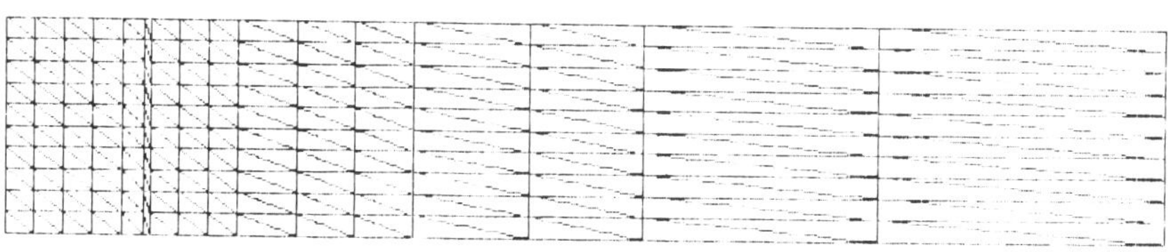

Figure 14. Finite-element idealization of the solution domain (Problem 1).

apparatus, which is deemed to be the physical equivalent of the problem described. For the numerical solution of this problem, the rectangular region shown in Figure 13, defined in the plane of the aquifer, is idealized by 320 elements and 187 nodes (Figure 14). Due to symmetry, only the half of the aquifer was modeled using the idealization shown. The idealization plot shown in Figure 14 was obtained from the IDEAL.BAS computer code.

On symmetry boundaries of the aquifer $\{x = 0 \text{ cm}, 0 \leq y \leq 50 \text{ cm}\}$ and $\{y = 0 \text{ cm}, y = 50 \text{ cm}, 0 \leq x \leq 200 \text{ cm}\}$, a Neumann condition with zero flux is defined, which in effect reduces the problem to a one-dimensional application. On the outer boundary $\{x = 200 \text{ cm}, 0 \leq y \leq 50 \text{ cm}\}$, a Dirichlet condition of constant head (H_o) is specified. The parameters in this application are defined as K = 0.42 cm/sec, S_y = 1.0, L = 200 cm, I = 0.056 cm/s, and H_o = 11.3 cm, which are the values used by Marino (1967). The finite-element results obtained for this problem are given comparatively with the analytical solution and the experimental results in Figure 15. Experimental results are shown for t = 9 minutes only in order to avoid crowding the presentation of results. The error criterion chosen in this application is ϵ = 1.0 E-06.

Numerical results obtained are in close agreement with the experimental results and the analytical solution outside the recharge zone; however, they are lower than the experimental values within the recharge zone. Similar deviations can also be seen between the analytical solution and the experimental results, which was attributed to observational errors and other effects, such as surface tension, in Marino's study. Analytical results, which are represented by solid lines in the same figure, are very close to numerical values. A discussion of this problem appears in Aral (1988). The solution time for this problem, on an IBM PS/2 microcomputer with an 8087 math coprocessor was 15 minutes. Piezometric head contour plots and piezometric head surface plots may also be obtained using the UPLOT1.DAT file generated. Figure 16 shows a typical topographic plot at time 480.0 seconds, and Figures 17 and 18 show surface plots at two different time levels.

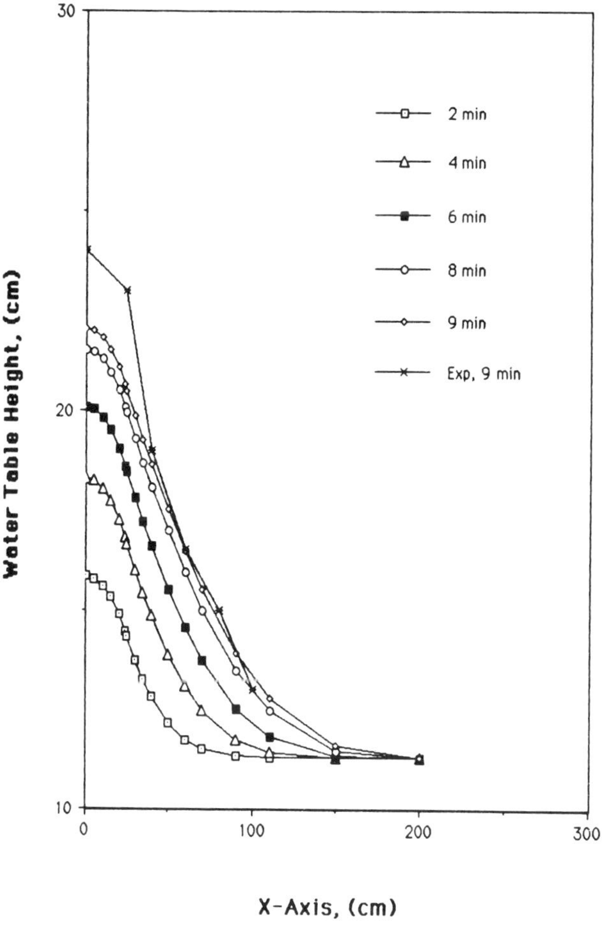

Figure 15. Comparison of numerical and analytical results (Problem 1).

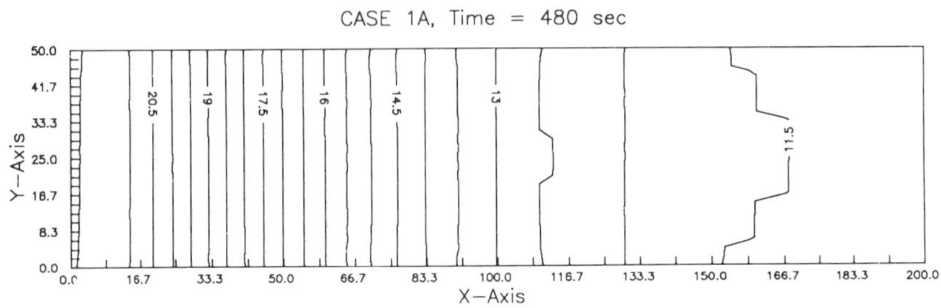

Figure 16. Piezometric head distribution in the main aquifer layer (time 480 sec).

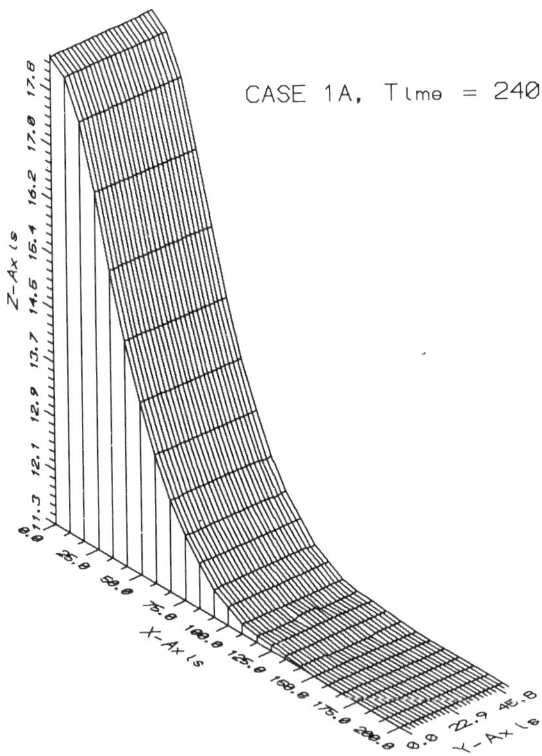

Figure 17. Piezometric head surface in the main aquifer layer (time 240 sec).

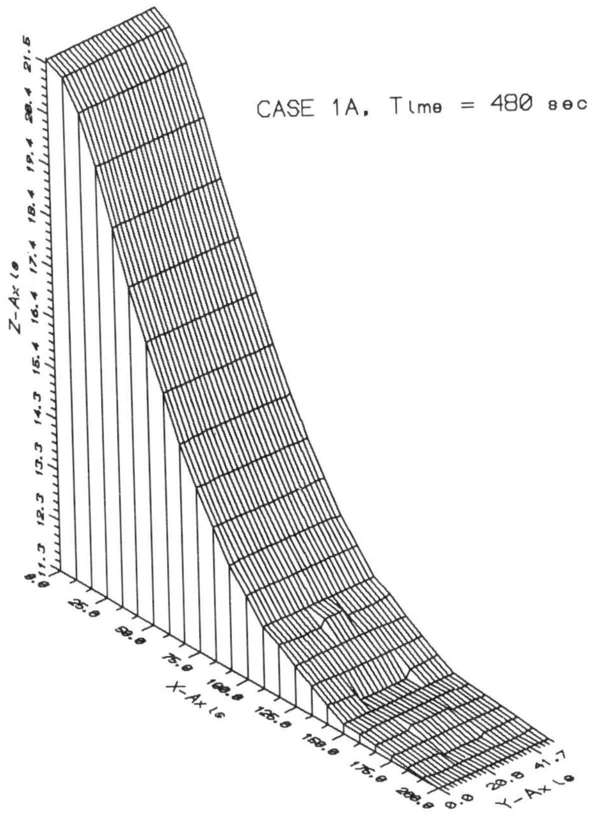

Figure 18. Piezometric head surface in the main aquifer layer (time 480 sec).

PROBLEM 2: TWO-DIMENSIONAL MOUND

A practical means of artificially recharging unconfined aquifers is through infiltration from surface water basins. With this procedure, the water is recharged over a large surface area, resulting in the growth and spread of a recharge mound. In this sense, this problem is similar to Problem 1, but the application is for a two-dimensional mound (Warner et al. 1989). In Warner's study, seven different analytical solutions to this problem were

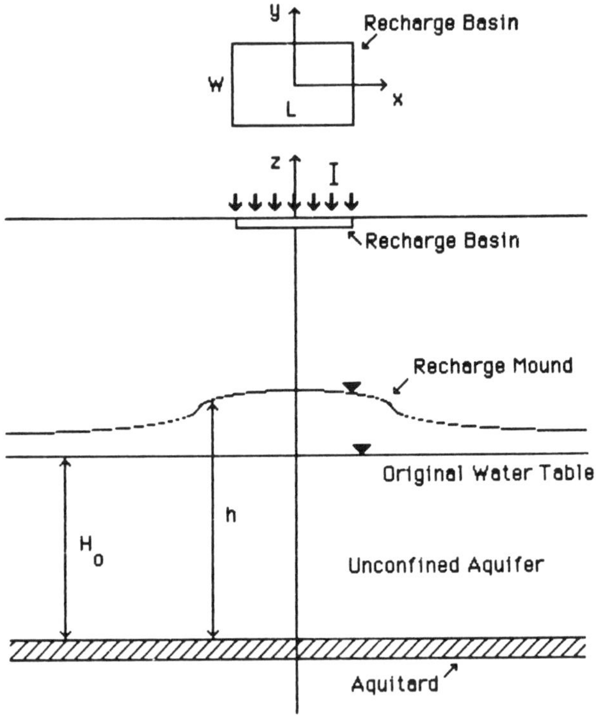

Figure 19. Definition sketch for Problem 2.

given, and solutions obtained from each analytic method were analyzed comparatively.

A definition sketch for Problem 2 is given in a cross section view in Figure 19. The characteristic field parameters used in this study, which are the same parameters used in Problem 1, are summarized in Table 1. Figure 20 shows the finite-element idealization of the solution domain, which again yielded 320 elements and 187 nodes. In this idealization, the element sizes chosen are different, but the idealization pattern is the same as for Problem 1. The error criterion chosen in this application was also the same as in the above problem, and only the upper right hand quadrant of the (x,y) domain, Figure 19, is modeled with symmetry axes chosen as positive (x) and positive (y) axis direc-

Table 1. Aquifer Data and Field Characteristics (Problem 2)

Recharge Rate, (I)	= 1.0 ft/days
Hydraulic Conductivity, (K)	= 10.0 ft/day
Initial Saturated Depth, (H_o)	= 200 ft
Specific Yield, (S_s)	= 0.2
Recharge Basin Width, (W)	= 200 ft
Recharge Basin Length, (L)	= 200 ft
Maximum Length in X-Dir.	= 16000 ft
Maximum Length in Y-Dir.	= 10000 ft

tions. The solution time on an IBM PS/2 microcomputer was again about 15 minutes.

The numerical results obtained for this problem are shown in Figure 21 for time 10, 20, 30, and 40 days. The difference between the numerical results and the analytical solution given by Warner et al. (1989) was less than 1%. Figures 22, 23, 24, and 25 summarize the contour and surface plots obtained at time 10 days and 30 days.

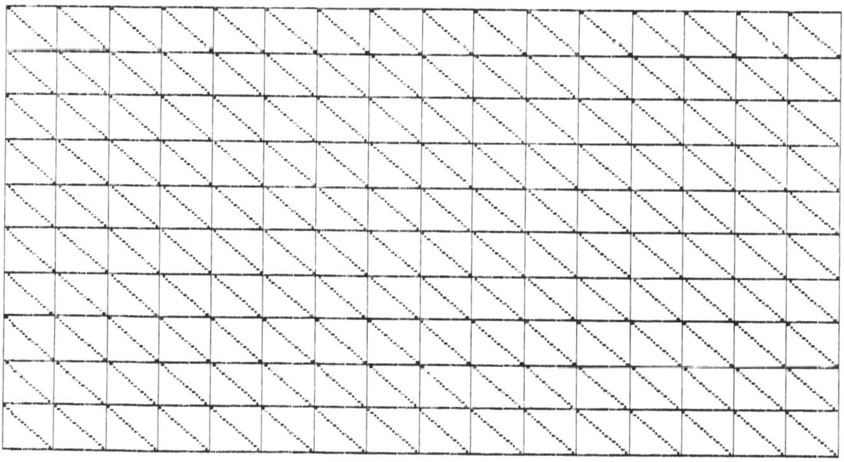

Figure 20. Finite-element idealization of solution domain (Problem 2).

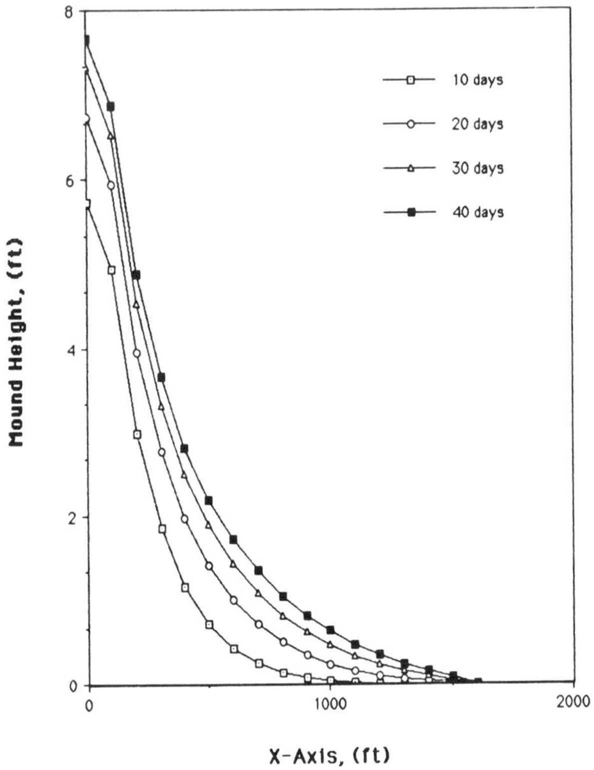

Figure 21. Comparison of numerical and analytical results (Problem 2).

PROBLEM 3: FIVE-LAYER SYSTEM

In this example (Aral 1986, 1987), a five-layer system is analyzed, and the numerical results are verified with the analytical solution given by Hunt (1985), who analyzed a system consisting of five semiconfined aquifers.

In Problem 3, the top aquitard has zero leakage; thus, the five-layer system is fully confined from above. The bottom aquifer has leakage from below, and an infinite aquifer system is modeled in the analytical solution. In the numerical solution, it is

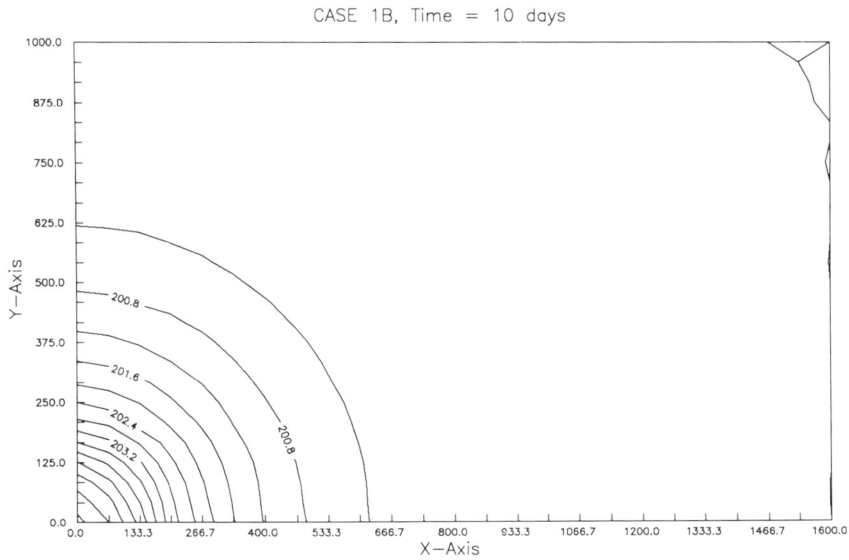

Figure 22. Piezometric head distribution at time 10 days (Problem 2).

assumed that the first aquifer layer is connected to an aquifer below that has a constant piezometric head distribution and that this piezometric head distribution does not vary in response to pumping operations performed in the second semi-confined aquifer layer. There is pumping in the second aquifer layer only.

The aquifer and aquitard parameters are given in Tables 2 and 3. The finite-element idealization chosen for this problem is shown in Figure 26, where the radius of the aquifer is chosen as 400 m as specified in Hunt's study, as opposed to the 900 m that was chosen earlier in the analysis of steady-state solutions in a five-layer aquifer (Aral 1990). Since the system has five layers in this application, the idealization shown in Figure 26 yields 585 nodes and 960 elements for the overall aquifer system analyzed.

In this application the solution required 22 minutes on an IBM PS/2 microcomputer with an 8087 math coprocessor. Numerical results obtained are summarized in Figures 27 and 28. In Figure 27, piezometric head values in each layer are

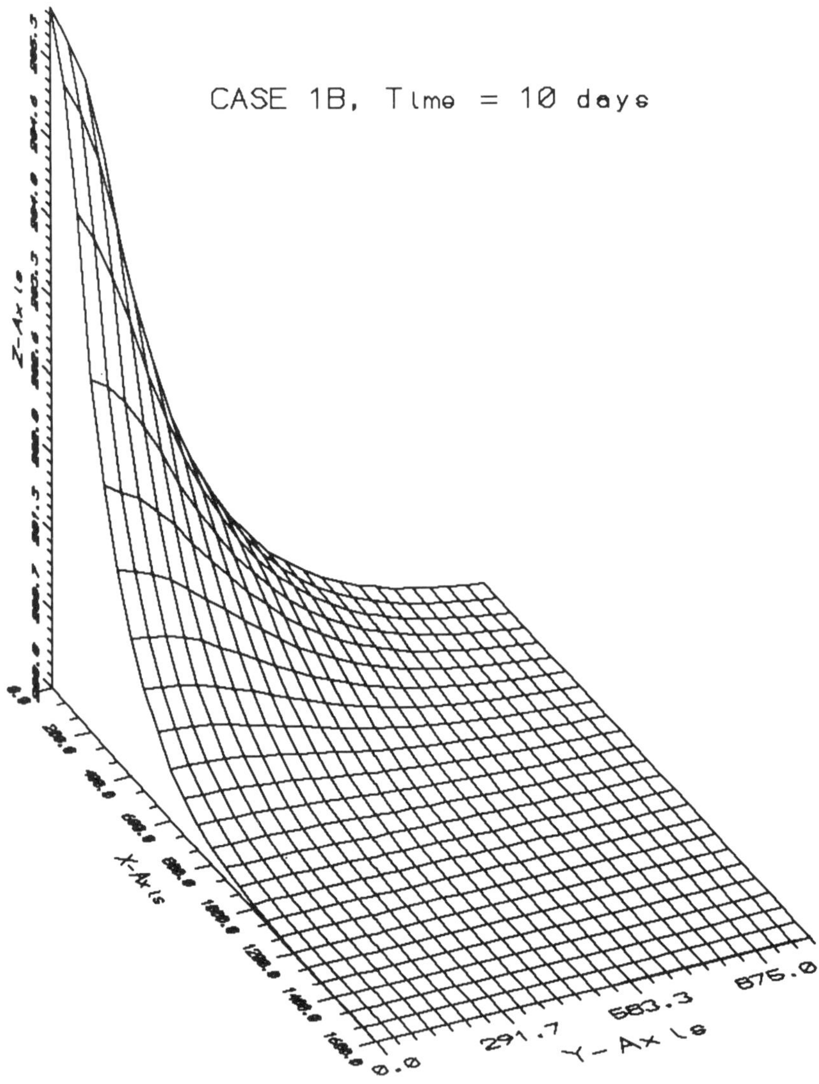

Figure 23. Piezometric head surface at time 10 days (Problem 2).

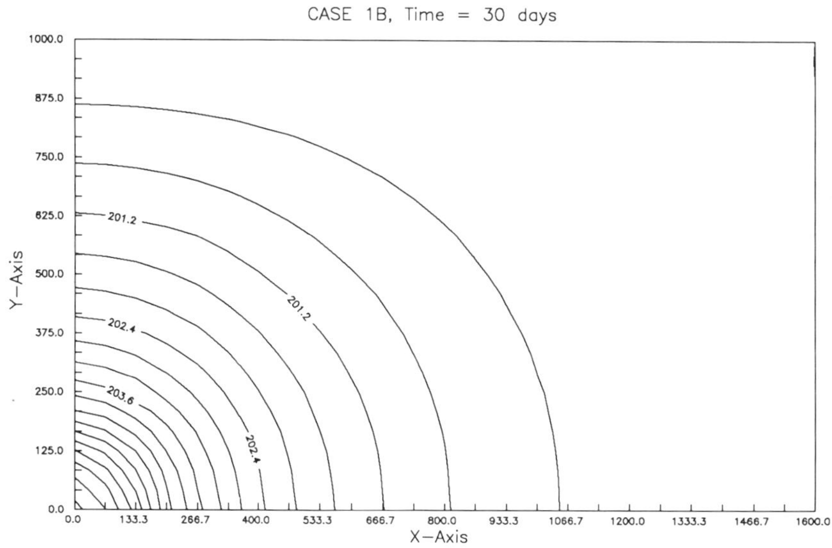

Figure 24. Piezometric head distribution at time 30 days (Problem 2).

shown as a function of time, and Figure 28 shows the plot of the piezometric head surface in each aquifer layer as the flow reaches steady-state conditions. The computation time includes the data generation phase for the five layers and the generation of the plot and discharge per unit width data files for the five layers. The convergence criterion was chosen as ϵ = 1.0 E-06, and the aquifer and aquitard layers are considered to be homogeneous and isotropic.

Numerical results shown in Figures 27 and 28 agree very closely with the analytical results given in the Hunt's study for all aquifer layers. The small differences in these results are expected and can be further reduced by choosing smaller elements near the well in all layers.

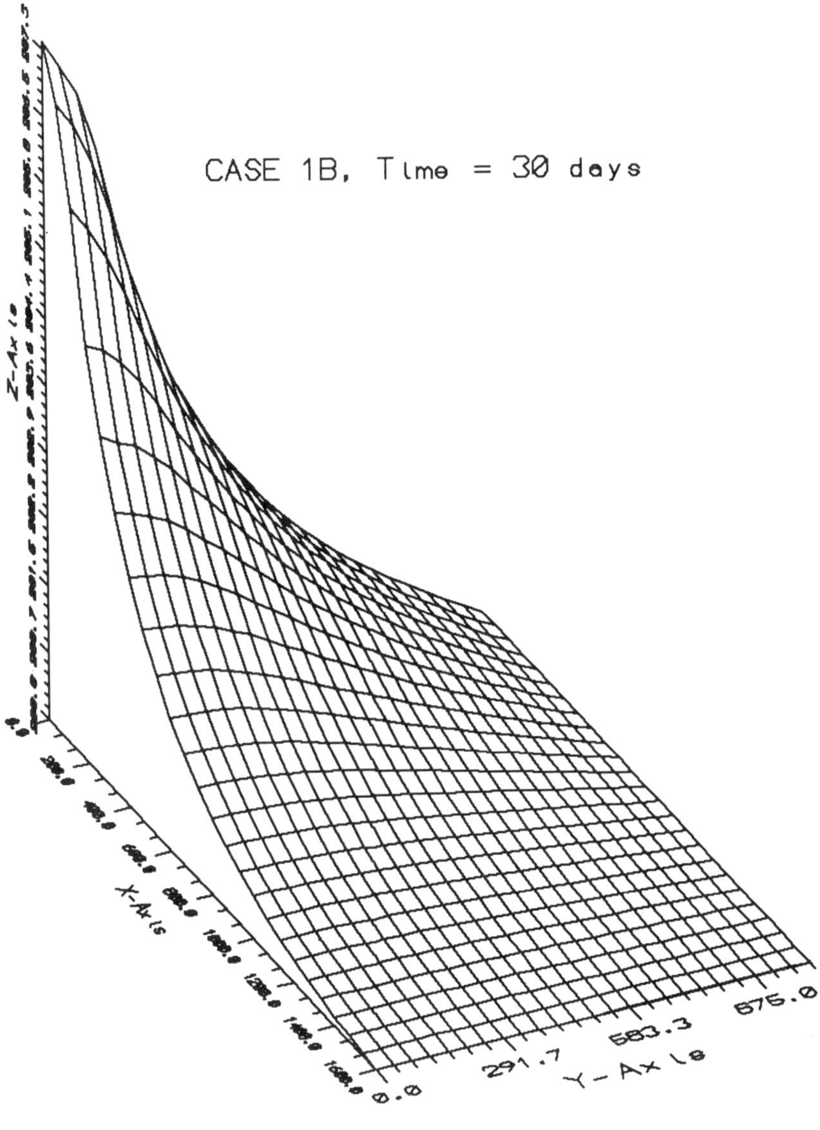

Figure 25. Piezometric head surface at time 30 days (Problem 2).

Table 2. Aquifer Data (Problem 3)

Aquifer #	Pumping Rate (m^3/sec)	Transmissivity (m^2/sec)	Storage Coeff.
1	0.0	0.001	0.0001
2	0.04	0.003	0.0002
3	0.0	0.0007	0.00008
4	0.0	0.0009	0.0004
5	0.0	0.0004	0.00005

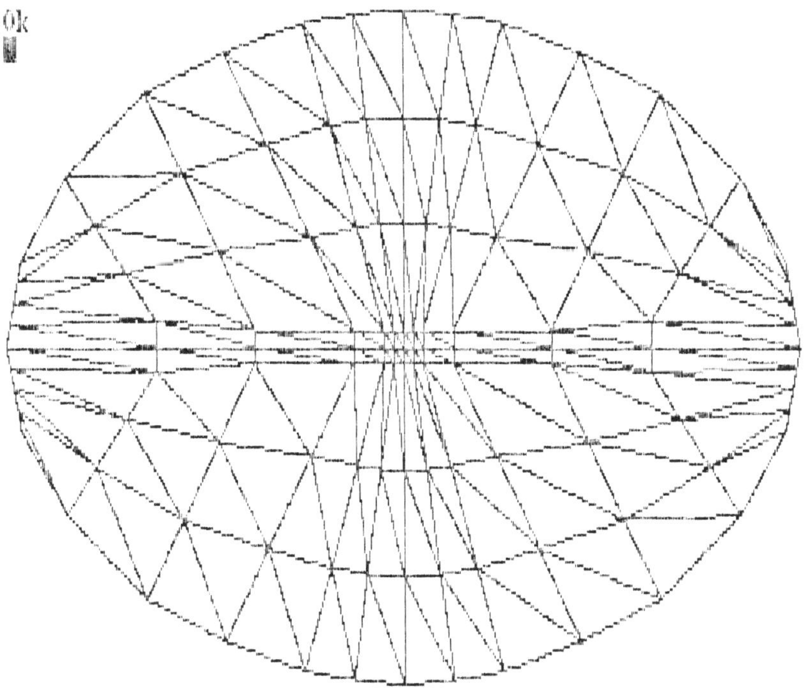

Figure 26. Finite-element idealization of Problem 3.

Table 3. Aquitard Data (Problem 3)

Aquitard #	K_z/d (1/sec)	Specific Storage (1/m)
1	0.00003	0.000
2	0.00001	0.000
3	0.000005	0.000
4	0.000008	0.000
5	0.000003	0.000
6	0.000	0.000

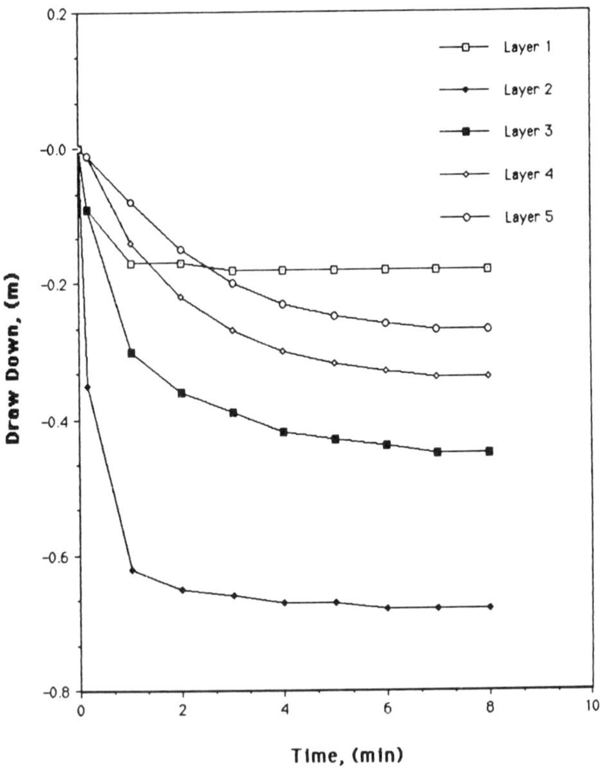

Figure 27. Drawdown for five layers as a function of time.

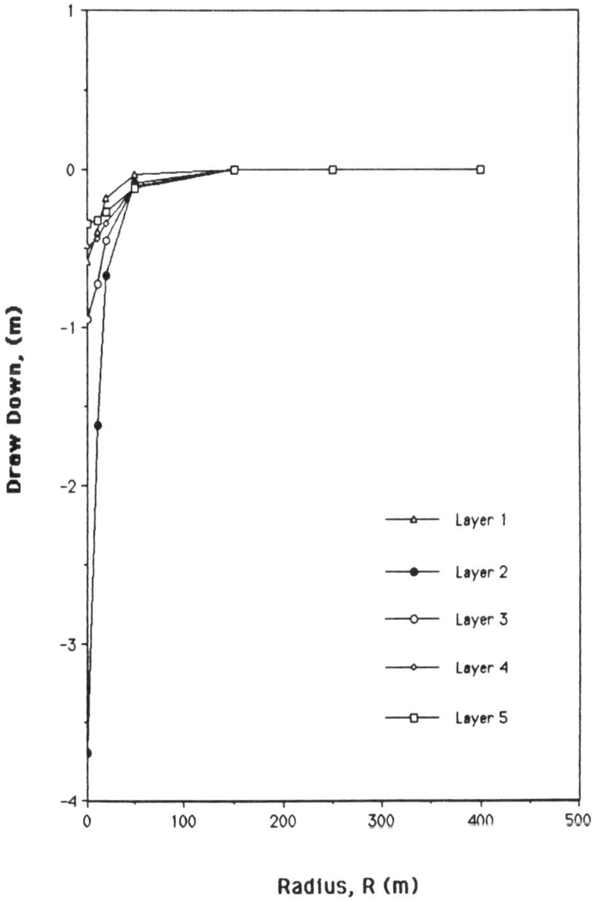

Figure 28. Steady-state results for the five-layer problem.

PROBLEM 4: MODEL GROUND WATER BASIN

As a final example, a model two-layer ground water basin is studied. In this application, nonhomogeneous aquifers and independent transient conditions for each layer are considered (Case 4B). The rectangular two-layer semiconfined ground water basin model is chosen as 16000 ft by 10000 ft in x and y directions, respectively (Figure 29). In this application, each layer of the basin is idealized by 320 right-angle triangular elements, which yields 187 nodes in each layer (Figure 30). Thus, in this application a total of 640 elements and 374 nodes were used.

Case 4A: Homogeneous, Isotropic System

In this case a homogeneous isotropic two-layer aquifer system is considered (Cheng and Ou 1989). The transmissivities of both layers are defined as 0.06962 ft^2/s, and K_z/d, for the leaky layer between them, is chosen as 7.7 E-07 1/s, where d is the thickness of the aquitard, K_z is the vertical permeability, and the two-layer aquifer system is assumed to be fully confined from the top and bottom. The specific storage of the middle aquitard is defined as 8.0 E-06 ft^{-1}. The storage coefficient of the main aquifer layers is chosen as 5.0 E-05. At boundaries parallel to y-direction at {x = 0 ft and x = 16000 ft} and also at the boundary {0 ≤ x ≤ 16000 ft, y = 10000 ft}, a Dirichlet boundary condition of H = 40 ft is specified, and the boundary parallel to x-direction at {y = 0} is defined as a no-flow boundary. The two-layer system is pumped from the first layer at position (x = 8000 ft, y = 3000 ft). The pumping rate is chosen as 0.4456 ft^3/s. The data file developed for this case is given in Appendix B as Case 4A data set.

Drawdown obtained at the well location for each layer is given comparatively in Figure 31. The contour and surface plots at t = 2 days are shown in Figures 32, 33, 34, and 35 for first and second layers. The computation time for this problem was about 32 minutes on an IBM PS/2 microcomputer.

Case 4B: Nonhomogeneous, Anisotropic System

In this case the data set prepared for Case 4A is modified to represent nonhomogeneous and anisotropic aquifers. Also, independent transient conditions are modeled in each aquifer layer. These conditions were specifically chosen to demonstrate the

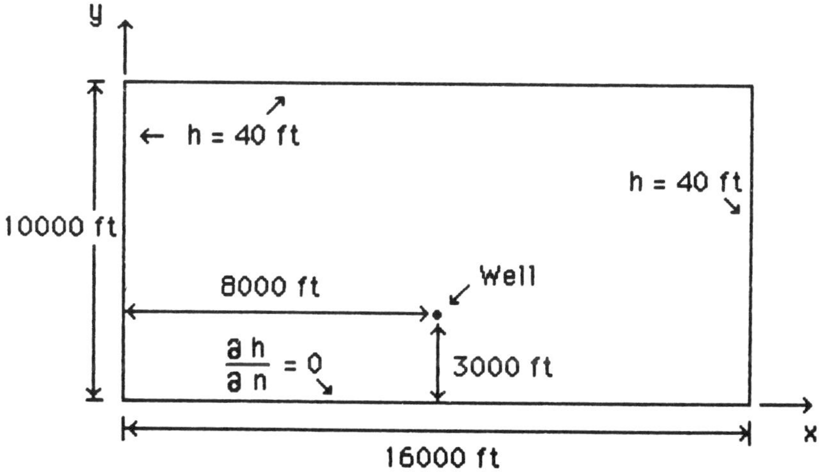

Figure 29. Definition sketch for Problem 4.

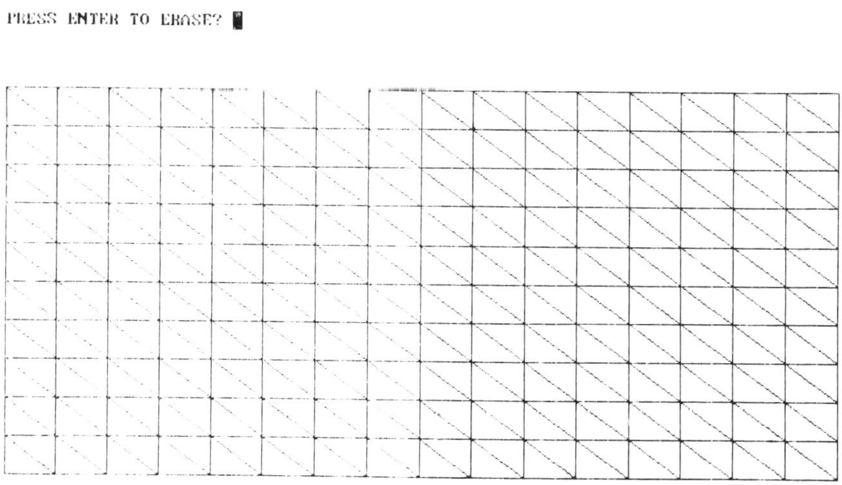

Figure 30. Finite-element idealization of Problem 4.

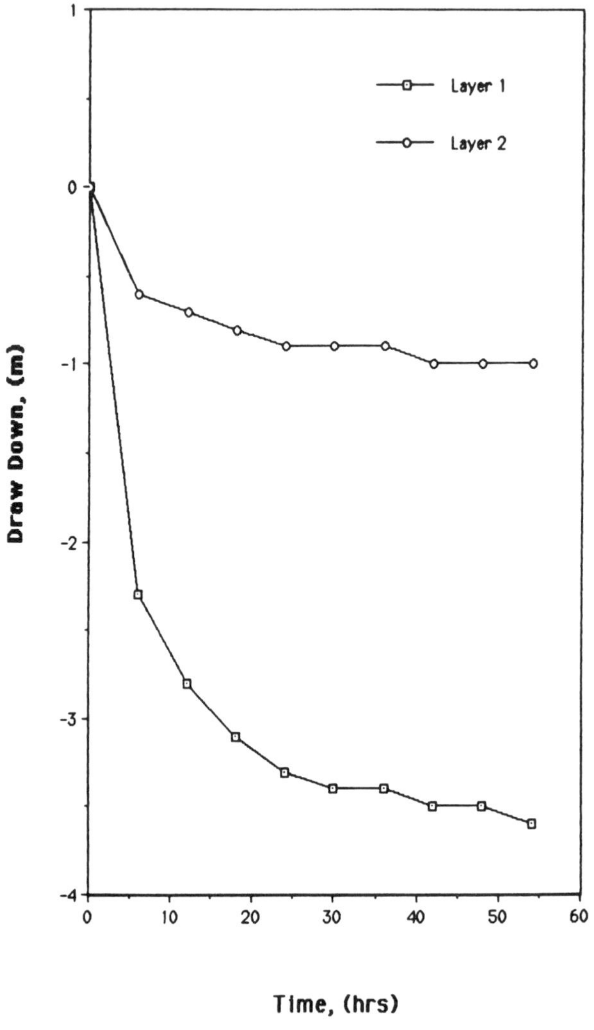

Figure 31. Comparison of drawdown at the well position.

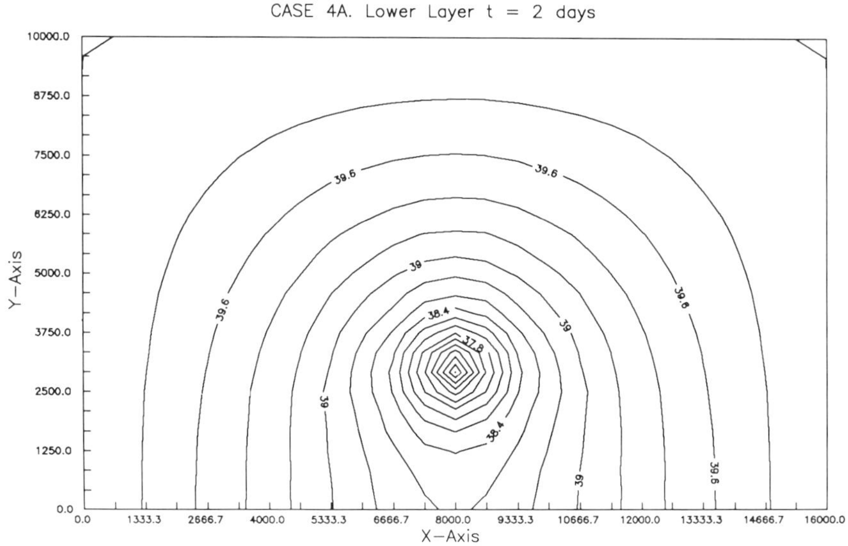

Figure 32. Piezometric head distribution in the first aquifer layer (Case 4A).

versatility of the computer code and to demonstrate the input data preparation required under such conditions.

The boundary conditions of this problem are as defined in Case 4A. The characteristic aquifer and aquitard properties used in this application are given in Tables 4 and 5. The pumping operations defined in Case 4A for the first layer are not changed, but an additional pump is started at time 0.5 days at position {x= 9000 ft, y = 5000 ft} in the second aquifer.

The data file for this application is given in Appendix B as Case 4B data set. The user should follow this data file and the procedures defined in Chapter 4 very carefully to generate data for nonhomogeneous aquifers with the variable pumping conditions defined above. Also, the data sets presented in the steady-state model (Aral 1990) should be studied carefully because similar data may be prepared for unsteady problems to analyze heterogeneous aquifers with time-dependent conditions. Both models basically use the same data generation format.

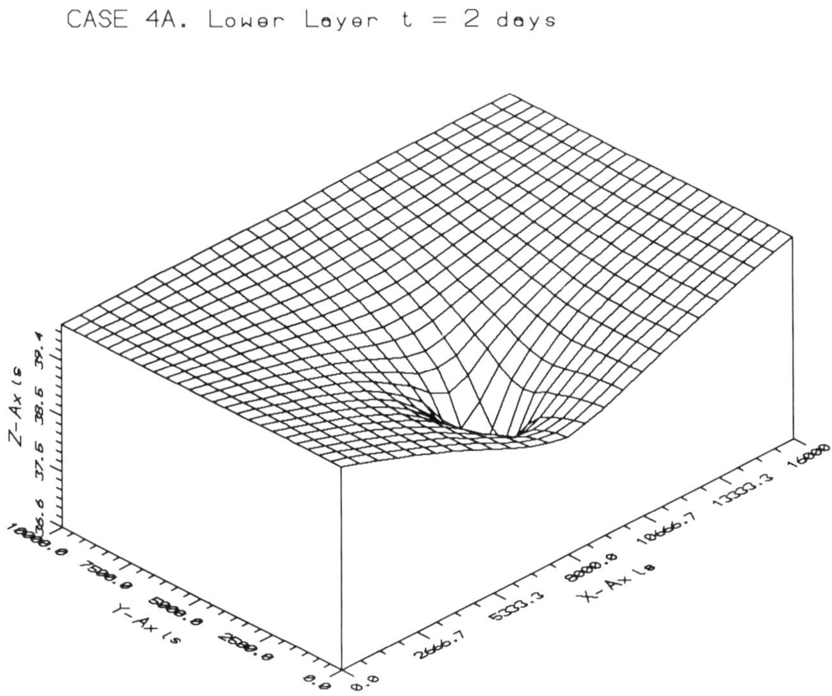

Figure 33. Piezometric head surface in the first aquifer layer (Case 4A).

Numerical results obtained in this application are again summarized as contour plot of piezometric head distribution in each aquifer layer and surface plot of piezometric head distribution in Figures 36, 37, 38, and 39 for first and second aquifer layers at t = 2 days. The computation time for this application was about 35 minutes.

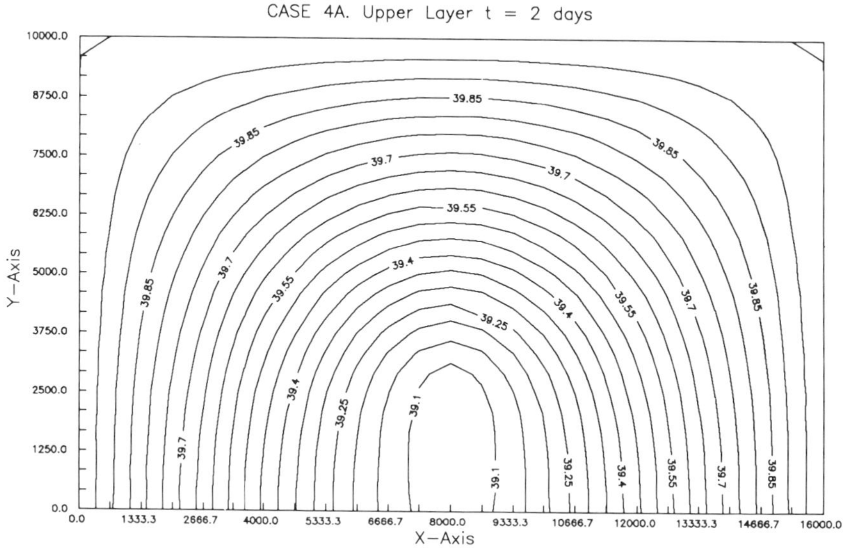

Figure 34. Piezometric head distribution in the second aquifer layer (Case 4A).

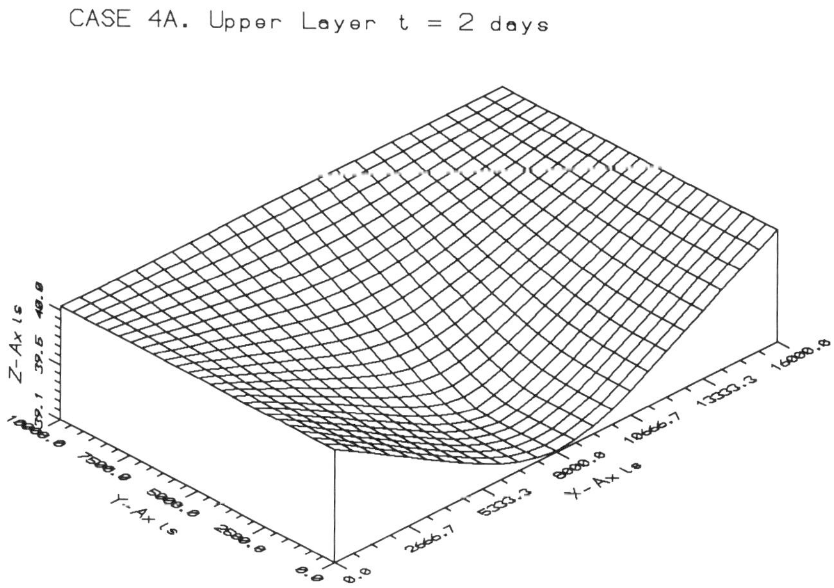

Figure 35. Piezometric head surface in the second aquifer layer (Case 4A).

Table 4. Aquifer Properties (Layer 1)

	T_x (ft^2/s)	T_y (ft^2/s)	Angle (Degrees)	S
$0 \le x \le 4000$ ft $0 \le y \le 10000$ ft	0.04	0.06	30	0.00004
$4000 \le x \le 16000$ ft $0 \le y \le 10000$ ft	0.06	0.08	40	0.00006

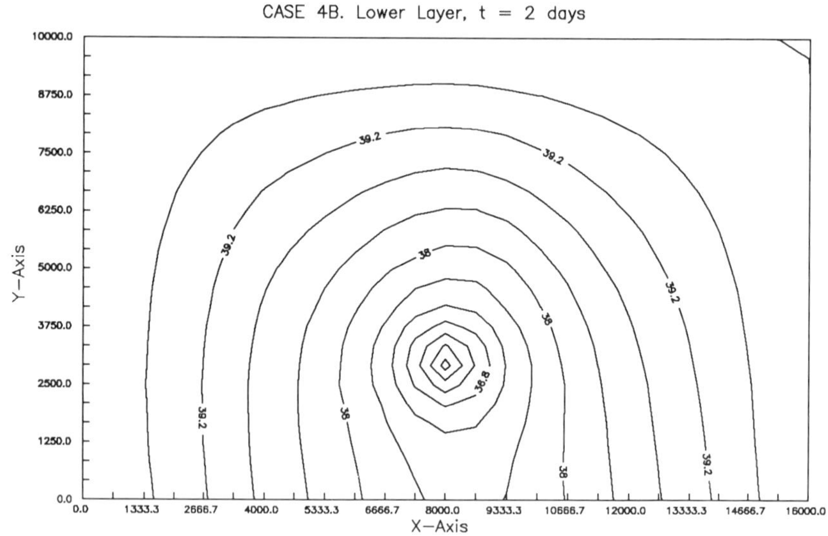

Figure 36. Piezometric head distribution in the first aquifer layer (Case 4B).

Table 5. Aquifer Properties (Layer 2)

	T_x (ft^2/s)	T_y (ft^2/s)	Angle (Degrees)	S
$0 \le x \le 10000$ ft $0 \le y \le 10000$ ft	0.01	0.05	10	0.00003
$10000 \le x \le 16000$ ft $0 \le y \le 10000$ ft	0.06	0.05	25	0.00006

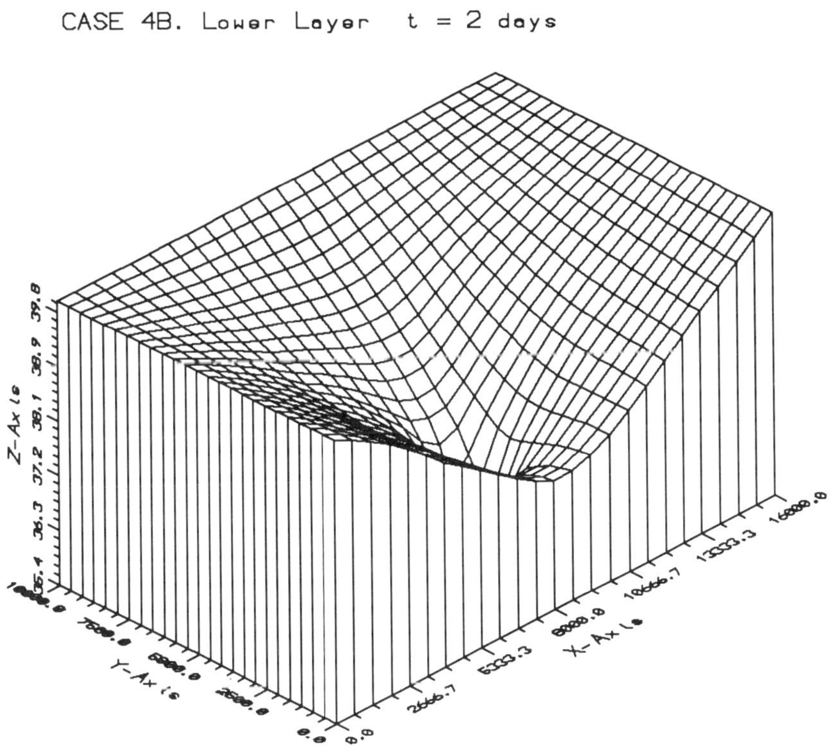

Figure 37. Piezometric head surface in the first aquifer layer (Case 4B).

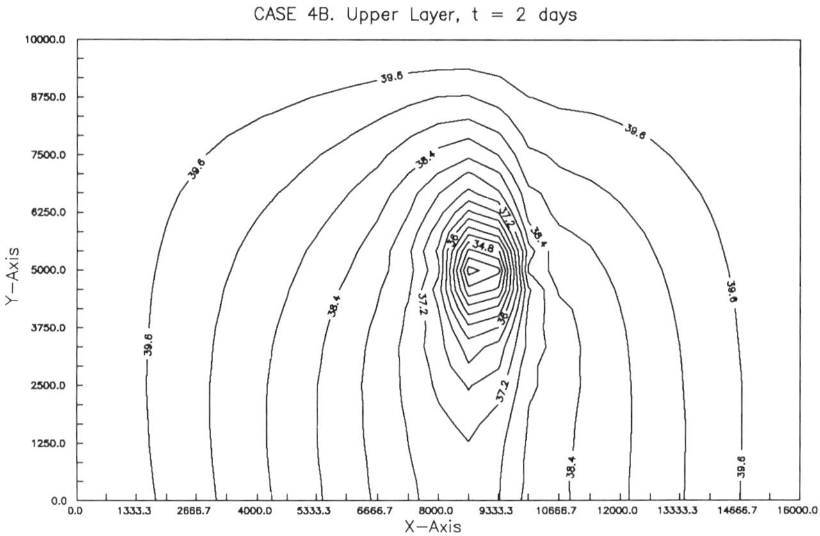

Figure 38. Piezometric head distribution in the second aquifer layer (Case 4B).

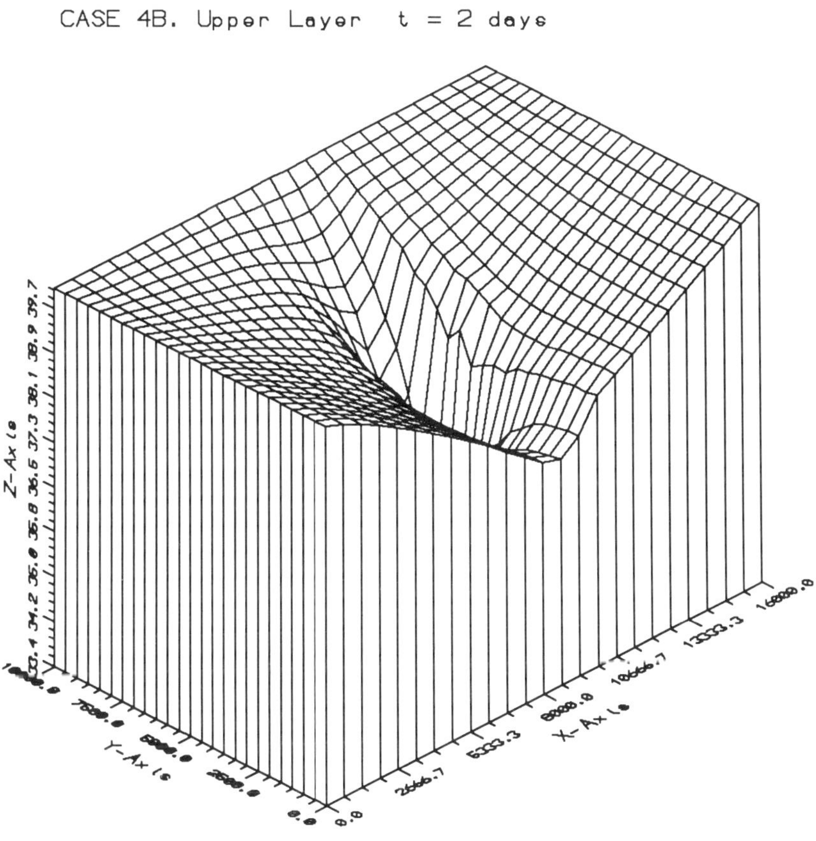

Figure 39. Piezometric head surface in the second aquifer layer (Case 4B).

Bibliography

Alexander, H.-D., C. Ou, and K. Ou. (1989). "An Efficient Laplace Transform Solution for Multiaquifer Systems," *Water Resour. Res.* 25(4):742–748.

Aral, M. M. (1985a). "A Simplified Approach to Regional Multilayered Aquifer Analysis," Environmental Resources Center, Georgia Institute of Technology, ERC 02-85, p. 65.

Aral, M. M. (1985b). "Aquifer Parameter Prediction in Leaky Aquifers," *J. Hydrol.* 80:19–44.

Aral, M. M. (1986). "A Regional Multilayered Aquifer Model for Microcomputers," *Microcomputers in Civil Eng.* 1(1):69–78.

Aral, M. M. (1989). "Waste Stabilization in Multilayer Aquifers by Optimal Hydraulic Control," *Groundwater J.* 27(4).

Aral, M. M. (1990). *Ground Water Modeling in Multilayer Aquifers: Steady Flow* (Chelsea, MI: Lewis Publishers, Inc.), p. 110.

Aral, M. M., and E. H. Haddad. (1987a). "Modeling Transient Ground Water Flow in Multilayered Aquifer Systems," Environmental Resources Center, Georgia Institute of Technology, ERC 05-86, p. 69.

Aral, M. M., and E. H. Haddad. (1987b). "An Unsteady Regional Aquifer Model for Microcomputers," *Microcomputers in Civil Eng.* 2(3):197–206.

Aral, M. M., and Y. Tang. (1988). "A New Boundary Element Formulation for Time-Dependent Confined and Unconfined Aquifer Problems," *Water Resour. Res.* 24(6):831–842.

Bear, J. (1972). *Dynamics of Fluids in Porous Media* (New York: Elsevier), p. 764.

Bear, J. (1979). *Hydraulics of Ground Water* (New York: McGraw-Hill Book Company), p. 569.

Bredehoeft, J. D., and G. F. Pinder. (1970). "Digital Analysis of Areal

Flow in Multiaquifer Ground Water Systems: A Quasi Three-Dimensional Model," *Water Resour. Res.* 6(3):883–888.

Carnahan, B. H., A. Luther, and J. O. Wilkes. (1969). *Applied Numerical Methods* (New York: John Wiley & Sons, Inc.).

Cook, R. D. (1974). *Concepts and Applications of Finite Element Analysis* (New York: John Wiley & Sons, Inc.), p. 402.

De Marsily, G. (1986). *Quantitative Hydrogeology* (Orlando, FL: Academic Press).

Frind, E. O., and J. M. Verge. (1978). "Three Dimensional Modeling of Groundwater Flow Systems," *Water Resour. Res.* 14(5):844–856.

Gray, W. G., and G. F. Pinder. (1976). "On the Relationship Between the Finite Element and Finite Difference Methods," *Int. J. Numer. Methods Eng.* 10:893–923.

Hantush, M. S. (1960). "Modification of the Theory of Leaky Aquifers," *J. Geophys. Res.* 65(11):3713–3725.

Hantush, M. S. (1964). "Hydraulics of Wells," in *Advances in Hydroscience, Vol. 1*, V. T. Chow, Ed. (New York: Academic Press, Inc.), pp. 281–442.

Hantush, M. S. (1967a). "Flow of Groundwater in Relatively Thick Leaky Aquifers," *Water Resour. Res.* 3(2):583.

Hantush, M. S. (1967b). "Flow to Wells in Aquifers Separated by a Semipervious Layer," *J. Geophys. Res.* 72(6):1709.

Hemker, C. J. (1984). "Steady Groundwater Flow in Leaky Multiple Aquifer Systems," *J. Hydrol.* 72:355–374.

Herrera, I. (1970). "Theory of Multiple Leaky Aquifers," *Water Resour. Res.* 6(1):185–193.

Hunt, B. (1985). "Flow to a Well in a Multiaquifer System," *Water Resour. Res.* 21(11):1637–1641.

Huyakorn, P. S., and G. F. Pinder. (1983). *Computational Methods in Subsurface Flow* (New York: Academic Press Inc.), p. 473.

Kinzelbach, W. (1986). *Groundwater Modeling: An Introduction with Sample Programs in BASIC*, Developments in Water Science 25 (Amsterdam: Elsevier), p. 333.

Lohman, S. W. (1972). "Definitions of Selected Ground-Water Terms – Revisions and Conceptual Refinements Geological Survey," U.S. Geological Survey Water Supply Paper 1988, p. 22.

Marino, M. A. (1967). "Hele-Shaw Model Study of the Growth and Decay of Ground Water Ridges," *J. Geophys. Res.* 72(4):1195–1205.

Meinzer, O. E. (1923). "Outline of Ground-Water Hydrology, with Definitions," U.S. Geological Survey Water Supply Paper 494, p. 71.

Neuman, S. P., C. Preller, and T. N. Narasimhan. (1982). "Adaptive

Explicit-Implicit Quasi Three-Dimensional Finite Element Model of Flow and Subsidence in Multiaquifer Systems," *Water Resour. Res.* 18(5):1551–1561.

Neuman, S. P., and P. A. Witherspoon. (1969a). "Theory of Flow in a Confined Two-Aquifer System," *Water Resour. Res.* 5(4):803–816.

Neuman, S. P., and P. A. Witherspoon. (1969b). "Applicability of Current Theories to Flow in Leaky Aquifers," *Water Resour. Res.* 5(4):817–829 (1969).

Pinder, G. F., and J. D. Bredehoeft. (1968). "Application of the Digital Computer for Aquifer Evaluation," *Water Resour. Res.* 4(5):1069–1093.

Pinder, G. F., and E. O. Frind. (1972). "Application of Galerkin's Procedure to Aquifer Analysis," *Water Resour. Res.* 8(1):108–120.

Pinder, G. F., and W. G. Gray. (1977). *Finite Element Simulation in Surface and Subsurface Hydrology* (New York: Academic Press, Inc.), p. 295.

Strack, O. D. L. (1989). *Groundwater Mechanics* (Englewood Cliffs, NJ: Prentice-Hall, Inc.), p. 732.

Varga, R. S. (1962). *Matrix Iterative Analysis* (Englewood Cliffs, NJ: Prentice-Hall, Inc.).

Verruijt, A. (1980). "Finite Element Calculations on a Microcomputer," *Int. J. Numer. Methods Eng.* 15:1570–1574 (1980).

Verruijt, A. (1982). *Theory of Groundwater Flow* (London: Macmillan).

Warner, J. W., D. Molden, M. Chehata, and D. K. Sunada. (1989). "Mathematical Analysis of Artificial Recharge from Basins," *Water Res. Bull.* 25(2):401–411.

Wilson, E. L., and R. W. Clough (1982). "Dynamic Response by Step-by-Step Matrix Analysis," Symposium on the Use of Computers in Civil Engineering, Lisbon.

Zienkiewicz, O. C. (1977). *The Finite Element Method* (London: McGraw-Hill).

Appendix A

Error Codes

The computer code ULAM.EXE is equipped with several run-time internal input data checking routines. These routines will catch most but not all possible errors that might be generated in a typical applications study. Though these routines will not check syntax errors in the data submitted or mixups in data order, they will control accuracy and compatibility of data, given the internal limits of the computer code. Syntax errors will appear in standard FORTRAN format if recognizable. These error messages will appear along with code numbers. A Microsoft FORTRAN manual will be helpful for further identification of these errors.

Most of the error messages listed below are run-time error messages and are self-explanatory. Detailed explanation of each error message is given whenever necessary. When an error occurs, these messages will be printed on the screen and execution will be stopped. In a typical run, the question mark place holders will be replaced by integers characteristic to the data analyzed.

1. **Number of elements cannot exceed 550**. Given the limits on dimensions assigned, number of elements in an aquifer layer may not exceed 550.

2. **Number of nodes in one layer cannot exceed 600**. Given the limits on dimensions, number of nodes in a layer cannot exceed 600. With this many nodes the system may only have one main aquifer layer.

3. **With the specified number of nodes the system can have a maximum of 1 layer**. More than one layer is specified with node number for each aquifer layer ranging between 301 and 600. This exceeds the limits specified. For a two-layer system the maximum number of nodes in a layer is 300.

4. **With the specified number of nodes, the system can have a maximum of 2 layers**. More than two layers are specified with node number for each layer ranging between 201 and 600. This exceeds the limits specified. For a three-layer system the maximum number of nodes in a layer is 200.

5. **With the specified number of nodes, the system can have a maximum of 3 layers**. More than three layers are specified with node number for each layer ranging between 151 and 600. This exceeds the limits specified. For a four-layer system the maximum number of nodes in a layer is 150.

6. **The system can have a maximum of 4 layers**. More than four layers are specified with node number for each layer ranging between 121 and 600. This exceeds the limits specified. For a five-layer system the maximum number of nodes in a layer is 120.

7. **Number of Dirichlet B. C. or Neumann B. C. cannot exceed 80 in each layer**. Dimension limits are exceeded.

8. **Number of wells in a layer cannot be > 30**. Dimension limits are exceeded.

9. **Number of Dirichlet B. C. cannot exceed 80 in each layer.** This statement implies that different numbers of Dirichlet boundary conditions are specified for each layer and that one of these specifications exceeds the dimension limits.

10. **Number of Neumann B. C. cannot exceed 80 in each layer.** This statement implies that different numbers of Neumann

boundary conditions are specified for each layer and that one of these specifications exceeds the dimension limits.

11. **Number of Head-Dependent B. C. cannot exceed 80 in each layer.** This statement implies that different numbers of head-dependent boundary conditions are specified for each layer and that one of these specifications exceeds the dimension limits.

12. **DIAGONAL ELEMENT IS NEGATIVE. NODAL POINT = ? LAYER = ? GIVES ERROR.** This error implies that a negative coefficient is computed at the diagonal position in the stiffness matrix. This is not possible in finite-element analysis. This error may occur if x,y coordinate data are not properly generated. Check input data.

13. **DIAGONAL ELEMENT IS ZERO. NODAL POINT = ? LAYER = ? GIVES ERROR.** Same as above, but in this case a zero value is encountered. Check input data.

14. **Error *** Upper band width > 20.** Upper band width limit is exceeded. Change your idealization. Upper band width is computed as the maximum difference between any two node numbers of an element plus 1.

15. **ERROR. Element area is zero, check x,y coordinate data set. Element No: ? gives error.** An element area is computed as zero. This error usually occurs when x,y coordinates of the nodes of the triangular elements are not generated properly. Check input data.

16. **ERROR. Piezometric head in an unconfined aquifer is zero or negative. Either excessive pumping or high confining layer permeabilities are specified, READ THE USER MANUAL. Layer number = ? Element Number = ?** This error usually occurs when a high vertical permeability value is assigned to a confining layer below an unconfined aquifer or the confining aquifer thickness is very small. Under such

conditions the pumping operations specified in the layer below the unconfined layer are basically draining the unconfined layer. The resistance to flow in the confining layer is low. This condition violates one of the assumptions specified in Chapter 2; namely, the difference between the permeabilities of confining layers and the main aquifer layers should be large, that is, the resistance to flow should be high. Otherwise, the confining layer soil matrix is not acting as a confining layer, which implies that the conceptual model for the multilayer aquifer system is not proper. Correct your input data set.

17. **ERROR. Confining aquifer layer thickness is zero. Layer number ?, Element number ?, gives error. Read the user manual! Assign Kz=0 if the confining layer is impervious.** For the element number and the layer number specified, confining aquifer thickness is zero. In the computer code this value cannot be assigned a zero value because it appears in the denominator of several equations (Chapter 2). In all applications the aquifer thickness should be specified as a positive value. However, a confining layer may be treated as an impervious layer by assigning $K_z = 0$ for an element.

18. **ERROR: ? nodes were generated. In generating X,Y-coordinates, generation exceeds the number of specified nodes. Check your input data.** This implies that nodes generated were more than the number of nodes specified on line 3 in input data. Check input data.

19. **ERROR: ? nodes were generated. In generating X,Y-coordinates, generation does not proceed through all specified nodes. Check your input data.** This implies that number of nodes generated is less than the number of nodes specified on line 3 in input data. Check input data.

20. **ERROR: In generating the connectivity matrix, ? elements were generated. Generation exceeds the specified number of**

elements. Check your input data. This implies that number of elements generated exceeds the number of elements specified on line 3 in input data. Check input data.

21. **ERROR: In generating the connectivity matrix, ? elements were generated. Generation does not proceed through all specified elements. Check your input data.** This implies that number of elements generated is less than the number of elements specified on line 3 in input data. Check input data.

22. **ERROR: In generating transmissivity data, ? elements were generated which does not match the specified number of elements. Check your input data.**

23. **ERROR: In generating conductivity data of the confining layers, ? elements were generated which does not match the specified number of elements. Check your input data.**

24. **ERROR: In generating thickness data of the confining layers, ? elements were generated which does not match the specified number of elements. Check your input data.**

25. **ERROR: In generating angle of anisotropy data, ? elements were generated which does not match the specified number of elements. Check your input data.**

26. **ERROR: In generating recharge data, ? elements were generated which does not match the specified number of elements. Check your input data.**

27. **ERROR: In generating base elevation data, ? nodes were generated which does not match the specified number of nodes. Check your input data.**

28. **ERROR: In generating upper and lower piezometric head values, ? nodes were generated which does not match the specified number of nodes. Check your input data.**

29. **ERROR: In generating specific storage coefficient data in the confining layers "?" elements were generated which does not match the specified number of elements. Check your input data.**

30. **ERROR: In generating storage coefficient data in the confined layers "?" elements were generated which does not match the specified number of elements. Check your input data.**

31. **Warning! ? terms are used in the computation of series. Results around element ?, layer ?, may not be as accurate as desired. Reduce TIME STEP!** This is an internal check on the number of terms used in the series, Equation 2.15, when time-dependent leakage is evaluated. The maximum number of terms used in this series is 1000 in the computer code. This warning will appear on the screen if the number of terms used exceeds 800. However, the execution will not be stopped, since the error introduced into the computation is in the order of $\epsilon = 0.5 \times 10^{-4}$. The user may decide to run the data again with a smaller time step and compare the two numerical results as a sensitivity analysis.

Appendix B

Input Data Files

In this appendix several data files used in Chapter 5 are given. These data files are also included on the distribution disk accompanying this text. Included in this appendix, but not the distribution disk, are explanations for each input data line to help the user follow the procedures given in Chapter 4 for data preparation. As shown below, explanations are given in brackets to the right of the data line. When preparing a data file, one should *not* include these explanation statements. The abbreviations used in these explanations, which refer to section headings given in Chapter 4, are as follows:

DGF = Data Generation Format, Chapter Four.
Item no. = Subsection of Data Generation Format, Chapter Four.

CASE 1 data file:

```
UFLW 1: ONE LYR. UNCONFINED AQUIFER  WITH RECHARGE.
                                                  [Title card: Item 1 DGF]
UNITS ARE IN CENTIMETERS, SECONDS                 [Units card: Item 1 DGF]
0, 11, 0, 0,187,320,50,0.000001, 0, 1, 1, 1,1,1,540.,3  [Control Variables: Item 2 DGF]
34,10,16        [Since data generation is selected above, reference nodes
                      and subintervals should be given here: Item 3 DGF]
1,0.,0.,1             [Reference node number, x-coordinate, y-coordinate
                              and generation control input: Item 4 DGF]
11,0.,50.,0                               [Same as above: Item 4 DGF]
12,5.,0.,1                                [Same as above: Item 4 DGF]
22,5.,50.,0                               [Same as above: Item 4 DGF]
23,10.,0.,1                               [Same as above: Item 4 DGF]
33,10.,50.,0                              [Same as above: Item 4 DGF]
34,15.,0.,1                               [Same as above: Item 4 DGF]
44,15.,50.,0                              [Same as above: Item 4 DGF]
45,20.,0.,1                               [Same as above: Item 4 DGF]
55,20.,50.,0                              [Same as above: Item 4 DGF]
56,23.8,0.,1                              [Same as above: Item 4 DGF]
66,23.8,50.,0                             [Same as above: Item 4 DGF]
67,25.,0.,1                               [Same as above: Item 4 DGF]
77,25.,50.,0                              [Same as above: Item 4 DGF]
78,30.,0.,1                               [Same as above: Item 4 DGF]
88,30.,50.,0                              [Same as above: Item 4 DGF]
89,35.,0.,1                               [Same as above: Item 4 DGF]
99,35.,50.,0                              [Same as above: Item 4 DGF]
100,40.,0.,1                              [Same as above: Item 4 DGF]
110,40.,50.,0                             [Same as above: Item 4 DGF]
111,50.,0.,1                              [Same as above: Item 4 DGF]
121,50.,50.,0                             [Same as above: Item 4 DGF]
122,60.,0.,1                              [Same as above: Item 4 DGF]
132,60.,50.,0                             [Same as above: Item 4 DGF]
133,70.,0.,1                              [Same as above: Item 4 DGF]
143,70.,50.,0                             [Same as above: Item 4 DGF]
144,90.,0.,1                              [Same as above: Item 4 DGF]
154,90.,50.,0                             [Same as above: Item 4 DGF]
155,110.,0.,1                             [Same as above: Item 4 DGF]
165,110.,50.,0                            [Same as above: Item 4 DGF]
166,150.,0.,1                             [Same as above: Item 4 DGF]
176,150.,50.,0                            [Same as above: Item 4 DGF]
177,200.,0.,1                             [Same as above: Item 4 DGF]
187,200.,50.,0                            [Same as above: Item 4 DGF]
319,.42,.42       [Hydraulic conductivity values of the first layer: Item 5 DGF]
319,0.        [Vertical permeability data for confining layer one: Item 6 DGF]
319,1.                       [Thickness of confining layer one: Item 7 DGF]
319,0.0                 [Specific storage of confining layer one: Item 8 DGF]
319,1.0         [Specific yield of the unconfined aquifer layer: Item 9 DGF]
319,0.0      [Angle of anisotropy of hydraulic conductivity values in aquifer
                                                layer one: Item 10 DGF]
319,0.0     [Initial recharge on the unconfined aquifer layer one: Item 11 DGF]
186,0.     [Elevation of the base of the unconfined aquifer layer: Item 12 DGF]
```

186,11.3 [Piezometric head at each node below the main aquifer layer: Item 13 DGF]
186,11.3 [Initial piezometric head in the unconfined aquifer layer: Item 13 DGF]
0 [Number of wells in layer one : Item 14 DGF]
177,11.3,178,11.3,184,11.3,179,11.3,180,11.3,181,11.3,182,11.3,183,11.3 [Dirichlet condition: Item 16a DGF]
185,11.3,186,11.3,187,11.3 [Dirichlet condition: Item 16a DGF]
10.,540.,12 [Time step, time limit on time step, printout interval: Item 17 DGF]
964000. [Time limit on the transients introduced on the analysis: Item 18a DGF]
0,1,0 [Control variables on the transients that will be introduced: Item 18b DGF]
99,0.056 [New infiltration rates for the unconfined aquifer layer: Item 18d DGF]
219,0.0 [New infiltration rates for the unconfined aquifer layer: Item 18d DGF]

CASE 2 data file:

```
CASE 2: ARTIFICIAL RECHARGE FROM BASINS.          [Title card: Item 1 DGF]
UNITS ARE IN FEET, DAYS                           [Units card: Item 1 DGF]
0, 27, 0, 0,187,320,50,0.000001, 0, 1, 1, 1,0,0,40.,3  [Control variables: Item 2 DGF]
34,10,16            [Since data generation is selected above, reference nodes
                    and subintervals should be given here: Item 3 DGF]
1,0.,0.,1           [Reference node number, x-coordinate, y-coordinate
                    and generation control input: Item 4 DGF]
11,0.,1000.,0       [Same as above: Item 4 DGF]
12,100.,0.,1        [Same as above: Item 4 DGF]
22,100.,1000.,0     [Same as above: Item 4 DGF]
23,200.,0.,1        [Same as above: Item 4 DGF]
33,200.,1000.,0     [Same as above: Item 4 DGF]
34,300.,0.,1        [Same as above: Item 4 DGF]
44,300.,1000.,0     [Same as above: Item 4 DGF]
45,400.,0.,1        [Same as above: Item 4 DGF]
55,400.,1000.,0     [Same as above: Item 4 DGF]
56,500.,0.,1        [Same as above: Item 4 DGF]
66,500.,1000.,0     [Same as above: Item 4 DGF]
67,600.,0.,1        [Same as above: Item 4 DGF]
77,600.,1000.,0     [Same as above: Item 4 DGF]
78,700.,0.,1        [Same as above: Item 4 DGF]
88,700.,1000.,0     [Same as above: Item 4 DGF]
89,800.,0.,1        [Same as above: Item 4 DGF]
99,800.,1000.,0     [Same as above: Item 4 DGF]
100,900.,0.,1       [Same as above: Item 4 DGF]
110,900.,1000.,0    [Same as above: Item 4 DGF]
111,1000.,0.,1      [Same as above: Item 4 DGF]
121,1000.,1000.,0   [Same as above: Item 4 DGF]
122,1100.,0.,1      [Same as above: Item 4 DGF]
132,1100.,1000.,0   [Same as above: Item 4 DGF]
133,1200.,0.,1      [Same as above: Item 4 DGF]
143,1200.,1000.,0   [Same as above: Item 4 DGF]
144,1300.,0.,1      [Same as above: Item 4 DGF]
154,1300.,1000.,0   [Same as above: Item 4 DGF]
155,1400.,0.,1      [Same as above: Item 4 DGF]
165,1400.,1000.,0   [Same as above: Item 4 DGF]
166,1500.,0.,1      [Same as above: Item 4 DGF]
176,1500.,1000.,0   [Same as above: Item 4 DGF]
177,1600.,0.,1      [Same as above: Item 4 DGF]
187,1600.,1000.,0   [Same as above: Item 4 DGF]
319,10.0,10.0       [Hydraulic conductivity values of the first aquifer layer: Item 5 DGF]
319,0.              [Vertical permeability data for confining layer one: Item 6 DGF]
319,1.              [Thickness of confining layer one: Item 7 DGF]
319,0.0             [Specific storage of confining layer one: Item 8 DGF]
319,0.2             [Specific yield of the unconfined aquifer layer: Item 9 DGF]
319,0.0             [Angle of anisotropy of hydraulic conductivity values in
                    unconfined aquifer layer: Item 10 DGF]
319,0.0             [Initial recharge data for the unconfined aquifer layer: Item 11 DGF]
186,0.              [Elevation of the base of the unconfined aquifer layer: Item 12 DGF]
186,200.0           [Piezometric head at each node below the main layer: Item 13 DGF]
```

186,200.0 [Initial piezometric head values of the unconfined aquifer layer: Item 13 DGF]
0 [Number of wells in layer one: Item 14 DGF]
11,200.,22,200.,33,200.,44,200.,55,200.,66,200.,77,200.,88,200.,99,200. [Dirichlet Condition: Item 16a DGF]
110,200.,121,200.,132,200.,143,200.,154,200.,165,200.,176,200. [Dirichlet Condition: Item 16a DGF]
177,200.,178,200.,184,200.,179,200.,180,200.,181,200.,182,200.,183,200. [Dirichlet Condition: Item 16a DGF]
185,200.,186,200.,187,200. [Dirichlet Condition: Item 16a DGF]
1.0,40.,10 [Time step, time limit on time step, printout interval: Item 17 DGF]
964000. [Time limit on the transients introduced on the analysis: Item 18a DGF]
0,1,0 [Control variables on the transients that will be introduced: Item 18b DGF]
0,1.0 [New infiltration rates for the unconfined aquifer layer: Item 18d DGF]
8,0.0 [New infiltration rates for the unconfined aquifer layer: Item 18d DGF]
0,1.0 [New infiltration rates for the unconfined aquifer layer: Item 18d DGF]
308,0.0 [New infiltration rates for the unconfined aquifer layer: Item 18d DGF]

CASE 3 data file:

FIVE LYR. AQUIFER: CIRC. DOMAIN, HUNT'S DATA. R = 900m [Title card: Item 1 DGF]
UNITS ARE IN METERS, SECONDS [Units card: Item 1 DGF]
5, 40, 0, 0,117,192,250,0.000001, 0, 1, 0, 1,0,1,440,3 [Control Variables: Item 2 DGF]
61,8,12 [Since data generation is selected above, reference nodes and subintervals should be given here: Item 3 DGF]
1,15.,300.,1 [Reference node number, x-coordinate, y-coordinate and generation control input: Item 4 DGF]
5,0.,400.,1 [Same as above: Item 4 DGF]
9,15.,500.,0 [Same as above: Item 4 DGF]
10,60.,200.,1 [Same as above: Item 4 DGF]
13,150.,370.,0 [Same as above: Item 4 DGF]
14,150.,400.,0 [Same as above: Item 4 DGF]
15,150.,430.,1 [Same as above: Item 4 DGF]
18,60.,600.,0 [Same as above: Item 4 DGF]
19,140.,100.,1 [Same as above: Item 4 DGF]
22,250.,380.,0 [Same as above: Item 4 DGF]
23,250.,400.,0 [Same as above: Item 4 DGF]
24,250.,420.,1 [Same as above: Item 4 DGF]
27,140.,700.,0 [Same as above: Item 4 DGF]
28,220.,50.,1 [Same as above: Item 4 DGF]
31,350.,380.,0 [Same as above: Item 4 DGF]
32,350.,400.,0 [Same as above: Item 4 DGF]
33,350.,420.,1 [Same as above: Item 4 DGF]
36,220.,750.,0 [Same as above: Item 4 DGF]
37,300.,15.,1 [Same as above: Item 4 DGF]
40,380.,380.,0 [Same as above: Item 4 DGF]
41,380.,400.,0 [Same as above: Item 4 DGF]
42,380.,420.,1 [Same as above: Item 4 DGF]
45,300.,785.,0 [Same as above: Item 4 DGF]
46,350.,5.,1 [Same as above: Item 4 DGF]
49,390.,380.,0 [Same as above: Item 4 DGF]
50,390.,400.,0 [Same as above: Item 4 DGF]
51,390.,420.,1 [Same as above: Item 4 DGF]
54,350.,795.,0 [Same as above: Item 4 DGF]
55,400.,0.,1 [Same as above: Item 4 DGF]
58,400.,380.,0 [Same as above: Item 4 DGF]
59,400.,400.,0 [Same as above: Item 4 DGF]
60,400.,420.,1 [Same as above: Item 4 DGF]
63,400.,800.,0 [Same as above: Item 4 DGF]
64,450.,5.,1 [Same as above: Item 4 DGF]
67,410.,380.,0 [Same as above: Item 4 DGF]
68,410.,400.,0 [Same as above: Item 4 DGF]
69,410.,420.,1 [Same as above: Item 4 DGF]
72,450.,795.,0 [Same as above: Item 4 DGF]
73,500.,15.,1 [Same as above: Item 4 DGF]
76,420.,380.,0 [Same as above: Item 4 DGF]
77,420.,400.,0 [Same as above: Item 4 DGF]
78,420.,420.,1 [Same as above: Item 4 DGF]
81,500.,785.,0 [Same as above: Item 4 DGF]

```
82,580.,50.,1             [Same as above: Item 4 DGF]
85,450.,380.,0            [Same as above: Item 4 DGF]
86,450.,400.,0            [Same as above: Item 4 DGF]
87,450.,420.,1            [Same as above: Item 4 DGF]
90,580.,750.,0            [Same as above: Item 4 DGF]
91,660.,100.,1            [Same as above: Item 4 DGF]
94,550.,380.,0            [Same as above: Item 4 DGF]
95,550.,400.,0            [Same as above: Item 4 DGF]
96,550.,420.,1            [Same as above: Item 4 DGF]
99,660.,700.,0            [Same as above: Item 4 DGF]
100,740.,200.,1           [Same as above: Item 4 DGF]
103,650.,370.,0           [Same as above: Item 4 DGF]
104,650.,400.,0           [Same as above: Item 4 DGF]
105,650.,430.,1           [Same as above: Item 4 DGF]
108,740.,600.,0           [Same as above: Item 4 DGF]
109,785.,300.,1           [Same as above: Item 4 DGF]
113,800.,400.,1           [Same as above: Item 4 DGF]
117,785.,500.,0           [Same as above: Item 4 DGF]
191,.001,.001             [Transmissivity values of the first layer: Item 5 DGF]
191,.003,.003             [Transmissivity values of the second layer: Item 5 DGF]
191,.0007,.0007           [Transmissivity values of the third layer: Item 5 DGF]
191,.0009,.0009           [Transmissivity values of the fourth layer: Item 5 DGF]
191,.0004,.0004           [Transmissivity values of the fifth layer: Item 5 DGF]
191,0.00003               [Vertical permeability data for confining layer one: Item 6 DGF]
191,0.00001               [Vertical permeability data for confining layer two: Item 6 DGF]
191,0.000005              [Vertical permeability data for confining layer three: Item 6 DGF]
191,0.000008              [Vertical permeability data for confining layer four: Item 6 DGF]
191,0.000003              [Vertical permeability data for confining layer five: Item 6 DGF]
191,0.                    [Vertical permeability data for confining layer six: Item 6 DGF]
191,1.                    [Thickness of confining layer one: Item 7 DGF]
191,1.                    [Thickness of confining layer two: Item 7 DGF]
191,1.                    [Thickness of confining layer three: Item 7 DGF]
191,1.                    [Thickness of confining layer four: Item 7 DGF]
191,1.                    [Thickness of confining layer five: Item 7 DGF]
191,1.                    [Thickness of confining layer six: Item 7 DGF]
191,0.0                   [Specific storage of confining layer one: Item 8 DGF]
191,0.0                   [Specific storage of confining layer two: Item 8 DGF]
191,0.0                   [Specific storage of confining layer three: Item 8 DGF]
191,0.0                   [Specific storage of confining layer four: Item 8 DGF]
191,0.0                   [Specific storage of confining layer five: Item 8 DGF]
191,0.0                   [Specific storage of confining layer six: Item 8 DGF]
191,.0001                 [Storage coefficient of aquifer layer one: Item 9 DGF]
191,.0002                 [Storage coefficient of aquifer layer two: Item 9 DGF]
191,.00008                [Storage coefficient of aquifer layer three: Item 9 DGF]
191,.0004                 [Storage coefficient of aquifer layer four: Item 9 DGF]
191,.00005                [Storage coefficient of aquifer layer five: Item 9 DGF]
191,0.0                   [Angle of anisotropy for aquifer layer one: Item 10 DGF]
191,0.0                   [Angle of anisotropy for aquifer layer two: Item 10 DGF]
191,0.0                   [Angle of anisotropy for aquifer layer three: Item 10 DGF]
191,0.0                   [Angle of anisotropy for aquifer layer four: Item 10 DGF]
191,0.0                   [Angle of anisotropy for aquifer layer five: Item 10 DGF]
116,10.                   [Piezometric head at node below the main aquifer layer: Item 13 DGF]
```

116,10. [Piezometric head at node above the main aquifer layer: Item 13 DGF]
0,0,0,0,0 [Number of wells in each layer: Item 14 DGF]
1,10.,2,10.,3,10.,4,10.,5,10.,6,10.,7,10.,8,10.,9,10.,10,10. [Dirichlet Condition: Item 16a DGF]
18,10.,19,10.,27,10.,28,10.,36,10.,37,10.,45,10.,46,10.,54,10. [Dirichlet Condition: Item 16a DGF]
55,10.,63,10.,64,10.,72,10.,73,10.,81,10.,82,10.,90,10. [Dirichlet Condition: Item 16a DGF]
91,10.,99,10.,100,10.,108,10.,109,10.,110,10.,111,10.,112,10. [Dirichlet Condition: Item 16a DGF]
113,10.,114,10.,115,10.,116,10.,117,10. [Dirichlet Condition: Item 16a DGF]
1,10.,2,10.,3,10.,4,10.,5,10.,6,10.,7,10.,8,10.,9,10.,10,10. [Dirichlet Condition: Item 16a DGF]
18,10.,19,10.,27,10.,28,10.,36,10.,37,10.,45,10.,46,10.,54,10. [Dirichlet Condition: Item 16a DGF]
55,10.,63,10.,64,10.,72,10.,73,10.,81,10.,82,10.,90,10. [Dirichlet Condition: Item 16a DGF]
91,10.,99,10.,100,10.,108,10.,109,10.,110,10.,111,10.,112,10. [Dirichlet Condition: Item 16a DGF]
113,10.,114,10.,115,10.,116,10.,117,10. [Dirichlet Condition: Item 16a DGF]
1,10.,2,10.,3,10.,4,10.,5,10.,6,10.,7,10.,8,10.,9,10.,10,10. [Dirichlet Condition: Item 16a DGF]
18,10.,19,10.,27,10.,28,10.,36,10.,37,10.,45,10.,46,10.,54,10. [Dirichlet Condition: Item 16a DGF]
55,10.,63,10.,64,10.,72,10.,73,10.,81,10.,82,10.,90,10. [Dirichlet Condition: Item 16a DGF]
91,10.,99,10.,100,10.,108,10.,109,10.,110,10.,111,10.,112,10. [Dirichlet Condition: Item 16a DGF]
113,10.,114,10.,115,10.,116,10.,117,10. [Dirichlet Condition: Item 16a DGF]
1,10.,2,10.,3,10.,4,10.,5,10.,6,10.,7,10.,8,10.,9,10.,10,10. [Dirichlet Condition: Item 16a DGF]
18,10.,19,10.,27,10.,28,10.,36,10.,37,10.,45,10.,46,10.,54,10. [Dirichlet Condition: Item 16a DGF]
55,10.,63,10.,64,10.,72,10.,73,10.,81,10.,82,10.,90,10. [Dirichlet Condition: Item 16a DGF]
91,10.,99,10.,100,10.,108,10.,109,10.,110,10.,111,10.,112,10. [Dirichlet Condition: Item 16a DGF]
113,10.,114,10.,115,10.,116,10.,117,10. [Dirichlet Condition: Item 16a DGF]
1,10.,2,10.,3,10.,4,10.,5,10.,6,10.,7,10.,8,10.,9,10.,10,10. [Dirichlet Condition: Item 16a DGF]
18,10.,19,10.,27,10.,28,10.,36,10.,37,10.,45,10.,46,10.,54,10. [Dirichlet Condition: Item 16a DGF]
55,10.,63,10.,64,10.,72,10.,73,10.,81,10.,82,10.,90,10. [Dirichlet Condition: Item 16a DGF]
91,10.,99,10.,100,10.,108,10.,109,10.,110,10.,111,10.,112,10. [Dirichlet Condition: Item 16a DGF]
113,10.,114,10.,115,10.,116,10.,117,10. [Dirichlet Condition: Item 16a DGF]
1.,10.,10 [Time step, time limit on time step, printout interval: Item 17 DGF]
500. [Time limit on the transients introduced on the analysis: Item 18a DGF]
1,0,0 [Control variables on the transients that will be introduced: Item 18b DGF]
0,1,0,0,0 [New number of wells in each layer: Item 18c DGF]

```
59           [New well node in the aquifer specified above: Item 18c DGF]
.04          [New well discharge in the aquifer specified above: Item 18c DGF]
5.,60.,10    [New time step, time limit on time step, printout interval: Item 17 DGF]
60.,440.,1   [New time step, time limit on time step, printout interval: Item 17 DGF]
```

CASE 4A data file:

```
UFLW 4A: TWO LYR. SYS. WITH A WELL FIELD,2 LYR. AQ. (16.E03FTx10.E03FT)
                                               [Title card: Item 1 DGF]
UNITS ARE IN FEET, SECONDS                     [Units card: Item 1 DGF]
2, 37, 0, 0,187,320,50,0.000001, 0, 1, 0, 1,0,1,180000.,3[Control variables: Item 2 DGF]
34,10,16             [Since data generation is selected above, reference nodes
                     and subintervals should be given here: Item 3 DGF]
1,0.,0.,1            [Reference node number, x-coordinate, y-coordinate
                     and generation control input: Item 4 DGF]
11,0.,10000.,0                           [Same as above: Item 4 DGF]
12,1000.,0.,1                            [Same as above: Item 4 DGF]
22,1000.,10000.,0                        [Same as above: Item 4 DGF]
23,2000.,0.,1                            [Same as above: Item 4 DGF]
33,2000.,10000.,0                        [Same as above: Item 4 DGF]
34,3000.,0.,1                            [Same as above: Item 4 DGF]
44,3000.,10000.,0                        [Same as above: Item 4 DGF]
45,4000.,0.,1                            [Same as above: Item 4 DGF]
55,4000.,10000.,0                        [Same as above: Item 4 DGF]
56,5000.,0.,1                            [Same as above: Item 4 DGF]
66,5000.,10000.,0                        [Same as above: Item 4 DGF]
67,6000.,0.,1                            [Same as above: Item 4 DGF]
77,6000.,10000.,0                        [Same as above: Item 4 DGF]
78,7000.,0.,1                            [Same as above: Item 4 DGF]
88,7000.,10000.,0                        [Same as above: Item 4 DGF]
89,8000.,0.,1                            [Same as above: Item 4 DGF]
99,8000.,10000.,0                        [Same as above: Item 4 DGF]
100,9000.,0.,1                           [Same as above: Item 4 DGF]
110,9000.,10000.,0                       [Same as above: Item 4 DGF]
111,10000.,0.,1                          [Same as above: Item 4 DGF]
121,10000.,10000.,0                      [Same as above: Item 4 DGF]
122,11000.,0.,1                          [Same as above: Item 4 DGF]
132,11000.,10000.,0                      [Same as above: Item 4 DGF]
133,12000.,0.,1                          [Same as above: Item 4 DGF]
143,12000.,10000.,0                      [Same as above: Item 4 DGF]
144,13000.,0.,1                          [Same as above: Item 4 DGF]
154,13000.,10000.,0                      [Same as above: Item 4 DGF]
155,14000.,0.,1                          [Same as above: Item 4 DGF]
165,14000.,10000.,0                      [Same as above: Item 4 DGF]
166,15000.,0.,1                          [Same as above: Item 4 DGF]
176,15000.,10000.,0                      [Same as above: Item 4 DGF]
177,16000.,0.,1                          [Same as above: Item 4 DGF]
187,16000.,10000.,0                      [Same as above: Item 4 DGF]
319,.06962,.06962       [Transmissivity values of the first layer: Item 5 DGF]
319,.06962,.06962      [Transmissivity values of the second layer: Item 5 DGF]
319,0.          [Vertical permeability data for confining layer one: Item 6 DGF]
319,0.0000007736 [Vertical permeability data for confining layer two: Item 6 DGF]
319,0.        [Vertical permeability data for confining layer three: Item 6 DGF]
319,1.                      [Thickness of confining layer one: Item 7 DGF]
319,100.                    [Thickness of confining layer two: Item 7 DGF]
319,1.                    [Thickness of confining layer three: Item 7 DGF]
319,0.              [Specific storage of confining layer one: Item 8 DGF]
```

```
319,0.000008      [Specific storage of confining layer two: Item 8 DGF]
319,0.            [Specific storage of confining layer three: Item 8 DGF]
319,0.00005       [Storage coefficient of aquifer layer one: Item 9 DGF]
319,0.00005       [Storage coefficient of aquifer layer two: Item 9 DGF]
319,0.     [Angle of anisotropy of transmissivity values in layer one: Item 10 DGF]
319,0.     [Angle of anisotropy of transmissivity values in layer two: Item 10 DGF]
186,40.    [Piezometric head values below main aquifer layers: Item 13 DGF]
186,40.    [Piezometric head values above main aquifer layers: Item 13 DGF]
0,0        [Number of wells in each layer: Item 14 DGF]
1,40.,2,40.,3,40.,4,40.,5,40.,6,40.,7,40.,8,40.,9,40.,10,40.   [Dirichlet condition:
                                                               Item 16a DGF]
11,40.,22,40.,33,40.,44,40.,55,40.,66,40.,77,40.,88,40.,99,40.,110,40.  [Dirichlet condition:
                                                               Item 16a DGF]
121,40.,132,40.,143,40.,154,40.,165,40.,176,40.,177,40.,178,40.,184,40.[Dirichlet condition:
                                                               Item 16a DGF]
179,40.,180,40.,181,40.,182,40.,183,40.,185,40.,186,40.,187,40.   [Dirichlet condition:
                                                               Item 16a DGF]
1,40.,2,40.,3,40.,4,40.,5,40.,6,40.,7,40.,8,40.,9,40.,10,40.   [Dirichlet condition:
                                                               Item 16a DGF]
11,40.,22,40.,33,40.,44,40.,55,40.,66,40.,77,40.,88,40.,99,40.,110,40.  [Dirichlet condition:
                                                               Item 16a DGF]
121,40.,132,40.,143,40.,154,40.,165,40.,176,40.,177,40.,178,40.,184,40.[Dirichlet condition:
                                                               Item 16a DGF]
179,40.,180,40.,181,40.,182,40.,183,40.,185,40.,186,40.,187,40.   [Dirichlet condition:
                                                               Item 16a DGF]
3600.,180000.,6  [Time step, time limit on time step, printout interval: Item 17 DGF]
964000.    [Time limit on the transients introduced on the analysis: Item 18a DGF]
1,0,0      [Control variables on the transients that will be introduced: Item 18b DGF]
1,0        [New number of wells in each aquifer layer: Item 18c DGF]
92         [New well position in aquifer layer one: Item 18c DGF]
.4456      [New well discharge in layer one: Item 18c DGF]
```

CASE 4B data file:

```
UFLW 4B: TWO LYR. SYS. WITH A WELL FIELD,2 LYR. AQ. (16.E03FTx10.E03FT)
                                              [Title card: Item 1 DGF]
UNITS ARE IN FEET, SECONDS                    [Units card: Item 1 DGF]
2, 37, 0, 0,187,320,50,0.000001, 0, 1, 0, 1,0,1,180000.,3[Control variables: Item 2 DGF]
34,10,16          [Since data generation is selected above, reference nodes
                  and subintervals should be given here: Item 3 DGF]
1,0.,0.,1         [Reference node number, x-coordinate, y-coordinate
                  and generation control input: Item 4 DGF]
11,0.,10000.,0                  [Same as above: Item 4 DGF]
12,1000.,0.,1                   [Same as above: Item 4 DGF]
22,1000.,10000.,0               [Same as above: Item 4 DGF]
23,2000.,0.,1                   [Same as above: Item 4 DGF]
33,2000.,10000.,0               [Same as above: Item 4 DGF]
34,3000.,0.,1                   [Same as above: Item 4 DGF]
44,3000.,10000.,0               [Same as above: Item 4 DGF]
45,4000.,0.,1                   [Same as above: Item 4 DGF]
55,4000.,10000.,0               [Same as above: Item 4 DGF]
56,5000.,0.,1                   [Same as above: Item 4 DGF]
66,5000.,10000.,0               [Same as above: Item 4 DGF]
67,6000.,0.,1                   [Same as above: Item 4 DGF]
77,6000.,10000.,0               [Same as above: Item 4 DGF]
78,7000.,0.,1                   [Same as above: Item 4 DGF]
88,7000.,10000.,0               [Same as above: Item 4 DGF]
89,8000.,0.,1                   [Same as above: Item 4 DGF]
99,8000.,10000.,0               [Same as above: Item 4 DGF]
100,9000.,0.,1                  [Same as above: Item 4 DGF]
110,9000.,10000.,0              [Same as above: Item 4 DGF]
111,10000.,0.,1                 [Same as above: Item 4 DGF]
121,10000.,10000.,0             [Same as above: Item 4 DGF]
122,11000.,0.,1                 [Same as above: Item 4 DGF]
132,11000.,10000.,0             [Same as above: Item 4 DGF]
133,12000.,0.,1                 [Same as above: Item 4 DGF]
143,12000.,10000.,0             [Same as above: Item 4 DGF]
144,13000.,0.,1                 [Same as above: Item 4 DGF]
154,13000.,10000.,0             [Same as above: Item 4 DGF]
155,14000.,0.,1                 [Same as above: Item 4 DGF]
165,14000.,10000.,0             [Same as above: Item 4 DGF]
166,15000.,0.,1                 [Same as above: Item 4 DGF]
176,15000.,10000.,0             [Same as above: Item 4 DGF]
177,16000.,0.,1                 [Same as above: Item 4 DGF]
187,16000.,10000.,0             [Same as above: Item 4 DGF]
79,.04,.06          [Transmissivity values of the first layer: Item 5 DGF]
239,.06,.08         [Transmissivity values of the first layer: Item 5 DGF]
199,.01,.05         [Transmissivity values of the second layer: Item 5 DGF]
119,.06,.05         [Transmissivity values of the second layer: Item 5 DGF]
319,0.              [Vertical permeability data for confining layer one: Item 6 DGF]
319,0.0000007736    [Vertical permeability data for confining layer two: Item 6 DGF]
319,0.              [Vertical permeability data for confining layer three: Item 6 DGF]
319,1.              [Thickness of confining layer one: Item 7 DGF]
319,100.            [Thickness of confining layer two: Item 7 DGF]
```

```
319,1.                               [Thickness of confining layer three: Item 7 DGF]
319,0.                               [Specific storage of confining layer one: Item 8 DGF]
319,0.000008                         [Specific storage of confining layer two: Item 8 DGF]
319,0.                               [Specific storage of confining layer three: Item 8 DGF]
79,0.00004                           [Storage coefficient of aquifer layer one: Item 9 DGF]
239,0.00006                          [Storage coefficient of aquifer layer one: Item 9 DGF]
199,0.00003                          [Storage coefficient of aquifer layer two: Item 9 DGF]
119,0.00006                          [Storage coefficient of aquifer layer two: Item 9 DGF]
79,30.     [Angle of anisotropy of transmissivity values in layer one: Item 10 DGF]
239,40.    [Angle of anisotropy of transmissivity values in layer one: Item 10 DGF]
199,10.    [Angle of anisotropy of transmissivity values in layer two: Item 10 DGF]
119,25.    [Angle of anisotropy of transmissivity values in layer two: Item 10 DGF]
186,40.          [Piezometric head values below main aquifer layers: Item 13 DGF]
186,40.          [Piezometric head values above main aquifer layers: Item 13 DGF]
0,0                               [Number of wells in each layer: Item 14 DGF]
1,40.,2,40.,3,40.,4,40.,5,40.,6,40.,7,40.,8,40.,9,40.,10,40.       [Dirichlet condition:
                                                                    Item 16a DGF]
11,40.,22,40.,33,40.,44,40.,55,40.,66,40.,77,40.,88,40.,99,40.,110,40.  [Dirichlet condition:
                                                                    Item 16a DGF]
121,40.,132,40.,143,40.,154,40.,165,40.,176,40.,177,40.,178,40.,184,40.[Dirichlet condition:
                                                                    Item 16a DGF]
179,40.,180,40.,181,40.,182,40.,183,40.,185,40.,186,40.,187,40.    [Dirichlet condition:
                                                                    Item 16a DGF]
1,40.,2,40.,3,40.,4,40.,5,40.,6,40.,7,40.,8,40.,9,40.,10,40.       [Dirichlet condition:
                                                                    Item 16a DGF]
11,40.,22,40.,33,40.,44,40.,55,40.,66,40.,77,40.,88,40.,99,40.,110,40.  [Dirichlet condition:
                                                                    Item 16a DGF]
121,40.,132,40.,143,40.,154,40.,165,40.,176,40.,177,40.,178,40.,184,40.[Dirichlet condition:
                                                                    Item 16a DGF]
179,40.,180,40.,181,40.,182,40.,183,40.,185,40.,186,40.,187,40.    [Dirichlet condition:
                                                                    Item 16a DGF]
3600.,180000.,6   [Time step, time limit on time step, printout interval: Item 17 DGF]
43200.       [Time limit on the transients introduced on the analysis: Item 18a DGF]
1,0,0      [Control variables on the transients that will be introduced: Item 18b DGF]
1,0                     [New number of wells in each aquifer layer: Item 18c DGF]
92                        [New well position in aquifer layer one: Item 18c DGF]
.4456                   [New well discharge for well in aquifer one: Item 18c DGF]
964000.      [Time limit on the transients introduced on the analysis: Item 18a DGF]
1,0,0      [Control variables on the transients that will be introduced: Item 18b DGF]
1,1                     [New number of wells in each aquifer layer: Item 18c DGF]
92                        [New well position in aquifer layer one: Item 18c DGF]
.4456                   [New well discharge for well in aquifer one: Item 18c DGF]
105                       [New well position in aquifer layer two: Item 18c DGF]
.4456                   [New well discharge for well in aquifer two: Item 18c DGF]
```

Appendix C

Two-Dimensional Regional Flow Equations

Equations 2.2 or 2.3 can be reduced to the form given in Equation 2.4 for confined aquifers or to the form given in Equation 2.5 for unconfined aquifers by integrating these equations in the vertical direction. In this process, vertically averaged values of all variables will replace the three-dimensional representations. In what follows, this vertical averaging process is summarized for a typical semiconfined aquifer, shown in Figure C-1. In this derivation, the main aquifer layer is assumed to be confined from above and below by leaky layers with a permeability much lower than that of the main aquifer. It is further assumed that only vertical flow exists in the confining layers, whereas the flow in the main layer is horizontal.

Given Figure C-1, one may define:

$$H(x,y,t) = z_2 - z_1, \qquad H \neq f(z) \tag{C-1}$$

Other relevant definitions may be given as: (1) average of a function:

$$\bar{f}(x,y,t) = \frac{1}{H} \int_{z_1}^{z_2} f(x,y,z,t)\, dz \tag{C-2}$$

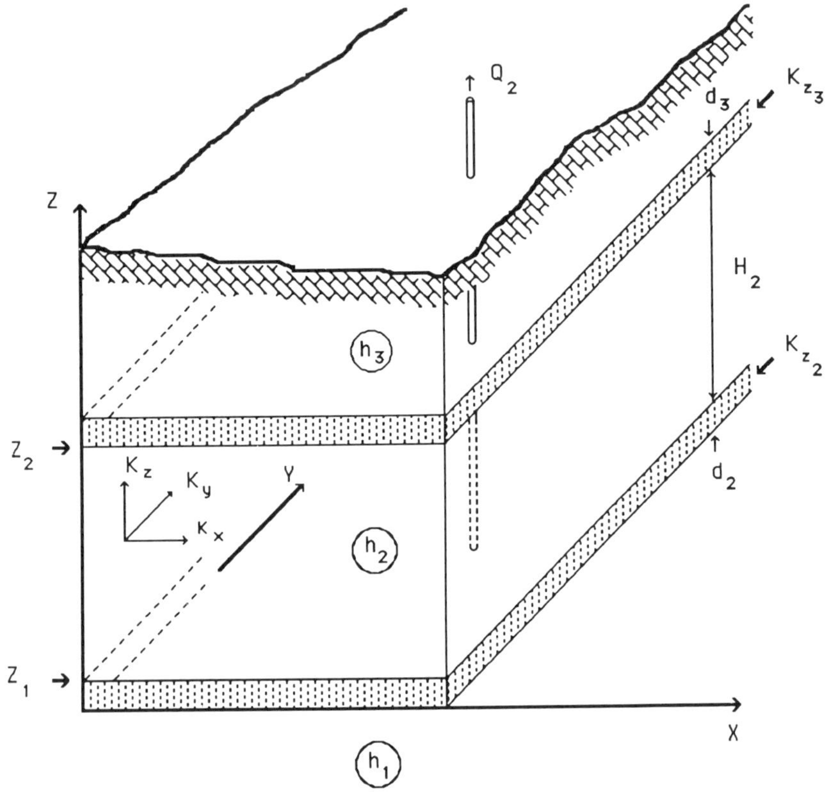

Figure C.1 Definition sketch for a semiconfined layer.

and (2) the Leibnitz rule:

$$\int_{z_1}^{z_2} \frac{\partial}{\partial x_i} f(x,y,z,t)\, dz = \frac{\partial}{\partial x_i} \int_{z_1}^{z_2} f(x,y,z,t)\, dz - f(x,y,z,t)\Big|_{z_2} \frac{\partial z_2}{\partial x_i} + f(x,y,z,t)\Big|_{z_1} \frac{\partial z_2}{\partial x_i} \qquad \text{(C-3)}$$

where $x_i = x, y, z, t$ for $i = 1, 2, 3, 4$. The three-dimensional ground water flow equation given earlier can be rewritten as:

$$S_s \frac{\partial h}{\partial t} + \frac{\partial q_x}{\partial x} + \frac{\partial q_y}{\partial y} + \frac{\partial q_z}{\partial z} = 0 \tag{C-4}$$

Each term of Equation C-4 can now be integrated in the z-direction yielding vertically averaged values.

$$\int_{z_1}^{z_2} \left(S_s \frac{\partial h}{\partial t} + \frac{\partial q_x}{\partial x} + \frac{\partial q_y}{\partial y} + \frac{\partial q_z}{\partial z} \right) dz = 0 \tag{C-5}$$

Using the Leibnitz rule, each term of the above equation can be written as:

$$\int_{z_1}^{z_2} \frac{\partial q_x}{\partial x} dz = \frac{\partial}{\partial x} \int_{z_1}^{z_2} q_x \, dz - q_x \Big|_{z_2} \frac{\partial z_2}{\partial x} + q_x \Big|_{z_1} \frac{\partial z_1}{\partial x} \tag{C-6}$$

$$\int_{z_1}^{z_2} \frac{\partial q_y}{\partial y} dz = \frac{\partial}{\partial y} \int_{z_1}^{z_2} q_y \, dz - q_y \Big|_{z_2} \frac{\partial z_2}{\partial y} + q_y \Big|_{z_1} \frac{\partial z_1}{\partial y} \tag{C-7}$$

$$\int_{z_1}^{z_2} \frac{\partial q_z}{\partial z} dz = q_z \Big|_{z_2} - q_z \Big|_{z_1} \tag{C-8}$$

$$S_s \int_{z_1}^{z_2} \frac{\partial h}{\partial t} dz = S_s \frac{\partial}{\partial t} \int_{z_1}^{z_2} h \, dz - S_s h \Big|_{z_2} \frac{\partial z_2}{\partial t} + S_s h \Big|_{z_1} \frac{\partial z_1}{\partial t} \tag{C-9}$$

Using the definition of an average of a function given in Equation C-2, one may substitute Equations C-6, C-7, C-8, and C-9 into Equation C-5. This process will yield:

$$S_s \frac{\partial (Hh)}{\partial t} + \frac{\partial}{\partial x}(Hq_x) + \frac{\partial}{\partial y}(Hq_y) + q_z \Big|_{z_2} - q_z \Big|_{z_1} - \mathbf{q}\Big|_{z_2} \cdot \nabla z_2$$

$$+ \mathbf{q}\Big|_{z_1} \cdot \nabla z_1 - S_s h \Big|_{z_2} \frac{\partial z_2}{\partial t} + S_s h \Big|_{z_1} \frac{\partial z_1}{\partial t} = 0 \quad \text{(C-10)}$$

where

$$\mathbf{q} = q_x \mathbf{i} + q_y \mathbf{j} \ ; \ \nabla = \frac{\partial}{\partial x}\mathbf{i} + \frac{\partial}{\partial y}\mathbf{j} \quad \text{(C-11)}$$

and q_x, q_y now represent vertically averaged values of Darcy velocities. The last two terms of Equation C-10 can be combined with the assumption that $h|_{z2} \approx h|_{z1} \approx h$, where h represents the average value of the piezometric head in the main aquifer layer.

$$-S_s h \left(\frac{\partial (z_2 - z_1)}{\partial t}\right) = -S_s h \frac{\partial H}{\partial t} \quad \text{(C-12)}$$

The first term of Equation C-10 can be expanded as:

$$S_s \frac{\partial (H\,h)}{\partial t} = S_s H \frac{\partial h}{\partial t} + S_s h \frac{\partial H}{\partial t} \quad \text{(C-13)}$$

Thus, the second term in Equation C-13 can be cancelled with the term generated in Equation C-12. With these substitutions, Equation C-10 can be written as:

$$S_s H \frac{\partial h}{\partial t} + \frac{\partial}{\partial x}(Hq_x) + \frac{\partial}{\partial y}(Hq_y) + q_z \Big|_{z_2} - q_z \Big|_{z_1}$$

$$- \mathbf{q}\Big|_{z_2} \cdot \nabla z_2 + \mathbf{q}\Big|_{z_1} \cdot \nabla z_1 = 0 \quad \text{(C-14)}$$

Mostly, one can assume that for regional aquifers the confining bed elevations (z_1, z_2) are spatially invariant. Thus, gradients of these terms will vanish except at well locations. Well bore is actually a point of discontinuity for z_2, but the following definition can be used for a discharging well to replace these gradient terms at the point of discontinuity.

$$\left(\mathbf{q}\Big|_{z_2} \cdot \nabla z_2\right)\Big|_{x_k,y_k} \doteq Q_w(x_k,y_k)\,\delta(x - x_k, y - y_k) \qquad \text{(C-15)}$$

Substituting these definitions into Equation C-14, one may obtain:

$$S_sH \frac{\partial h}{\partial t} + \frac{\partial}{\partial x}(Hq_x) + \frac{\partial}{\partial y}(Hq_y) - Q_w + q_z\Big|_{z_2} - q_z\Big|_{z_1} = 0 \qquad \text{(C-16)}$$

Substituting the definition of Darcy's law, one may obtain:

$$S_sH \frac{\partial h}{\partial t} - \frac{\partial}{\partial x}\left(HK_x \frac{\partial h}{\partial x}\right) - \frac{\partial}{\partial y}\left(HK_y \frac{\partial h}{\partial y}\right) - Q_w$$

$$+ q_z\Big|_{z_2} - q_z\Big|_{z_1} = 0 \qquad \text{(C-17)}$$

where ($HK_x=T_x$ and ($HK_y=T_y$) represent the transmissivities of the main aquifer layer, and ($S_sH=S$) represents the storage coefficient. For unconfined aquifers this product is identified as specific yield, S_y, or sometimes as effective porosity. The leakage terms, q_z, will be replaced by the unsteady analytical representation of leakage between two layers, using the definition of Darcy's law in z-direction for aquitards and the analytical solution of the governing partial differential equation, as shown in Chapter 2.

Appendix D

IDEAL.BAS Computer Code

The computer program given below may be used to plot the data file IDEAL.DAT, generated by the computer code ULAM.EXE, using a format similar to that used in the steady-state solution (Aral 1990). The data file generated is already in a proper format and is ready to be read from the program IDEAL.BAS. The program should be run through a BASIC or BASICA compiler. The screen size chosen in the program below is 640 by 480. For screens with other resolutions, statements 360, 400, and 480 (below) should be changed before executing the program.

When the program is properly executed, messages will be printed on the screen to facilitate proper input data sequences. The first input required is the input data file name, which is IDEAL.DAT unless the file name has been changed. The second input required is the specification of the disk drive from which the data file may be retrieved. The user may specify disk drives A:\, B:\, or C:\. Then, at the prompt, ENTER must be pressed to start the plotting of the idealization on the screen. Hard copy of this plot may be obtained by pressing PRINT SCREEN after the screen plotting is completed, if a compatible printer is available.

The program given below may also be compiled under Microsoft QuickBASIC compiler, in which case the command SCREEN 9 (line 470) may be chosen for high-resolution screens, if available. In such cases, the color command may also be changed for various color choices. Under these new conditions the idealization figures will be plotted with much higher resolution, and the execution time will be much shorter. This program is also included on the distribution disk and should be run on compatible microcomputers which have Q-BASIC compilers. In particular, the code BRUN20.EXE is needed to execute this program. BRUN20.EXE file is a Q-BASIC compiler program. If Q-BASIC compiler is not available, then IDEAL.BAS computer code should be used with a standard basic compiler.

```
 10 REM PROGRAM IDEAL.BAS
 20 CLS : KEY OFF
 30 REM
 40 PRINT " PLOT OF THE IDEALIZED DOMAIN USING TRIANGULAR
    ELEMENTS "
 50 PRINT : PRINT : INPUT " DATA FILE NAME ";N$
 60 PRINT: PRINT: INPUT " DATA IS ON DISK DRIVE = A:/B:/C: ";C$
 70 IF C$ = "B:" GO TO 110
 80 IF C$ = "C:" GO TO 130
 90 N$ = "A:" + N$
100 GO TO 140
110 N$ = "B:" + N$
120 GO TO 140
130 N$ = "C:" + N$
140 OPEN N$ FOR INPUT AS #1
150 INPUT #1, NLAYR1 'NUMBER OF LAYERS
160 INPUT #1, N  'NUMBER OF NODES
170 INPUT #1, M  'NUMBER OF ELEMENTS
180 PRINT : PRINT "  NUMBER OF LAYERS   = ";NLAYR1
190 PRINT : PRINT "  NUMBER OF NODES    = ";N
200 PRINT : PRINT "  NUMBER OF ELEMENTS = ";M
210 DIM X(N),Y(N),IN%(M,3)
220 FOR I = 1 TO N
230 INPUT #1, X(I),Y(I)
240 NEXT I
250 FOR I = 1 TO M
260 INPUT #1, IN%(I,1),IN%(I,2),IN%(I,3)
270 NEXT I
280 XMAX = - 1E+20 : XMIN = 1E+20
290 YMAX = - 1E+20 : YMIN = 1E+20
300 FOR I = 1 TO N
310 IF X(I) < XMIN THEN XMIN = X(I)
320 IF X(I) > XMAX THEN XMAX = X(I)
330 IF Y(I) < YMIN THEN YMIN = Y(I)
340 IF Y(I) > YMAX THEN YMAX = Y(I)
350 NEXT I
360 SX = 640 / (XMAX-XMIN) : SY = 480 / (YMAX-YMIN)
370 SC = SY : IF SX < SY THEN SC = SX
380 FOR I = 1 TO N
390 X(I) = (X(I) - XMIN) * SC
400 Y(I) = 480 - (Y(I) - YMIN) * SC
410 NEXT I
420 CLOSE #1
430 BEEP
440 PRINT : PRINT "WHAT FOLLOWS IS THE IDEALIZATION YOU HAVE
    CHOSEN"
450 INPUT "PRESS ENTER TO CONTINUE ";P$
460 REM DRAWING OF THE IDEALIZATION
470 CLS : KEY OFF: SCREEN 2:
480 WINDOW SCREEN (0,0) - (640,480)
490 HCOLOR = 1
500 FOR I = 1 TO M
```

```
510 FOR J = 1 TO 3
520 N2 = J+1 : IF J = 3 THEN N2 = 1
530 X1 = X(IN%(I,J)) : Y1 = Y(IN%(I,J))
540 X2 = X(IN%(I,N2)): Y2 = Y(IN%(I,N2))
550 GOSUB 580
560 NEXT J,I
570 END
580 REM SUBROUTINE BEGIN
590 LE = (X1 - X2)^2 + (Y1 - Y2)^2
600 LE = SQR(LE)
610 NE = INT(LE)
620 FOR IDOT = 1 TO NE
630 XP = X1 + (IDOT - 1) * (X2-X1)/NE
640 YP = Y1 + (IDOT - 1) * (Y2-Y1)/NE
650 XP = INT(XP + .1) : YP = INT(YP + .1)
660 PSET(XP,YP),HCOLOR
670 NEXT IDOT
680 RETURN
690 REM SUBROUTINE END
```

Appendix E

Glossary

In two supplementary texts, *Ground Water Modeling in Multilayer Aquifers: Steady Flow* (Aral 1990) and the present text, steady and unsteady ground water flow models in multilayer aquifers are presented. These models may be utilized by personnel from different disciplines, such as engineers, geologists, environmentalists and scientists. To assist in better communication between these interdisciplinary fields, a glossary of the most important terms used in the field of geohydrology is provided (Lohman 1972; Meinzer 1923):

Aquifer – A formation, group of formations, or part of a formation that contains sufficient saturated permeable material to yield significant quantities of water to wells and springs. The term *aquifer* was defined by O. E. Meinzer from a geological concept in which water bodies are classified in accordance with stratigraphy or rock types. Meinzer (1945, p. 159) clearly intended that an aquifer include the unsaturated part of the permeable unit.

Artesian – Synonymous with *confined. Artesian water* and *artesian water body* are equivalent, respectively, to *confined* ground water and *confined* water body. An *artesian well* is a well deriving its water from an artesian or *confined* water body. The water level in an artesian well stands above the top of the artesian water body it taps. If the water level in an artesian well stands above the land surface, the well is a *flowing artesian well.*

Aquiclude – A saturated formation that is incapable of transmitting significant quantities of water under ordinary field conditions.

Aquifuge – An impervious formation that neither contains nor transmits water.

Aquitard – A saturated formation that transmits low quantities of water under ordinary field conditions.

Area of influence of a well (L^2) – The area surrounding a well within which the piezometric surface has been lowered when pumping has produced the maximum steady rate of flow.

Confined aquifer – An aquifer that contains water under pressure significantly greater than atmospheric, and its upper limit is the bottom of a bed of distinctly lower hydraulic conductivity than the aquifer. Since the water is under pressure, water levels in wells rise above the top of the aquifer. *Artesian aquifer* is a commonly used synonym. Confined aquifers are classified as *leaky* or *nonleaky* depending upon whether ground water in the aquifer is confined by aquitards or aquicludes.

Capillary fringe – The zone immediately above the water table in which all or some of the interstices are filled with water that is under less than atmospheric pressure and that is continuous with the water below the water table. The water is held above the water table by interfacial forces (surface tension). The capillary fringe is typically saturated to some distance above its base at the water table; upward from the saturated part only progressively smaller pores are filled, and the upper limit is indistinct. In some quantitative studies it is convenient to define the upper limit as the level at which 50% of the pore space is filled with water.

Some lateral flow generally occurs throughout the capillary fringe, but because the effective hydraulic conductivity decreases rapidly with moisture content, the lateral flow in the capillary fringe is generally negligible compared with that in the saturated zone. The thickness of the capillary fringe is greater in fine-grained material than in coarse-grained material. Consequently, where the water table intersects materials of different particle size, vertical components of movement are superimposed on lateral movement of water through the capillary fringe.

Confining bed – A body of less permeable material overlying or underlying an aquifer. It is a saturated formation or group of formations that yield a very low rate of water compared to an aquifer. It is semipervious in nature; however, over a large horizontal area, it may permit the passage of large amounts of water between adjacent aquifers, which it lies between.

Conductivity, hydraulic, K (LT^{-1}) – Used interchangeably with the term *field coefficient of permeability*. If a porous medium is isotropic and the fluid is homogeneous, the *hydraulic conductivity* of the medium is the volume of water at the existing kinematic viscosity that will move in unit time under a unit hydraulic gradient through a unit area measured at right angles to the direction of flow. It is a property of the fluid and the medium.

Darcy's law can be expressed as:

$$q = \frac{Q}{A} = -K\frac{dh}{dl} \qquad \text{(E-1)}$$

K, the constant of proportionality in Equation E-1, is the hydraulic conductivity and may be expressed as:

$$K = k\rho g/\mu \qquad \text{(E-2)}$$

where ρ is density of water, g is the gravitational constant, μ is the dynamic viscosity of water, and k is identified as intrinsic permeability.

In anisotropic media the direction of the specific discharge (q) is not generally parallel to that of the gradient (dh/dl) of the head. In such media the cartesian components of the specific discharge are related to those of the gradient by:

$$\begin{aligned} -q_x &= K_{xx}\frac{\partial h}{\partial x} + K_{xy}\frac{\partial h}{\partial y} + K_{xz}\frac{\partial h}{\partial z}, \\ -q_y &= K_{yx}\frac{\partial h}{\partial x} + K_{yy}\frac{\partial h}{\partial y} + K_{yz}\frac{\partial h}{\partial z}, \\ -q_z &= K_{zx}\frac{\partial h}{\partial x} + K_{zy}\frac{\partial h}{\partial y} + K_{zz}\frac{\partial h}{\partial z} \end{aligned} \tag{E-3}$$

The quantities in the form K_{xx}, K_{xy}, K_{xz}, and so forth, called *conductivity coefficients,* are the components of a second-order tensor, generally symmetric. The physical meaning of the conductivity coefficients may be visualized from the following examples: K_{xx} is a conductivity coefficient along direction x that contributes a component of flux along the x-axis due to the component of head gradient along x $(=\partial h/\partial x)$. K_{zx} is a conductivity coefficient along direction z that contributes a component of flux along the z-axis due to the component of head gradient along x $(=\partial h/\partial x)$. K_{xz} is a conductivity coefficient along direction x that contributes a component of flux along the x-axis due to the component of head gradient along the z axis $(=\partial h/\partial z)$.

Cone of depression – The depression, in the shape of an inverted cone, of the piezometric surface of a body of groundwater which defines the area of influence of a well.

Drawdown – The vertical distance the free water elevation is lowered or the reduction of the pressure head due to the removal of free water.

Flow, steady – Flow occurring when at *any* point the magnitude and direction of the specific discharge are constant in time (*see also* Flow, unsteady).

Flow, uniform – Flow occurring when at *every* point the specific discharge has the same magnitude and direction. A property is uniform if, at a given instant, it is the same at every point.

Flow, unsteady—Flow occurring when at *any* point the magnitude or direction of the specific discharge changes with time (*see also* Flow, steady).

The word *transient* is used in reference to the temporary features of unsteady flow. Thus, in unsteady flow, the specific discharge, the piezometric head, and perhaps other factors consist of a steady component plus a transient component.

Ground water, confined—Ground water that is under pressure significantly greater than atmospheric, and its upper limit is the bottom of a bed of distinctly lower hydraulic conductivity than that of the material in which the confined water occurs.

Ground water, perched—Unconfined ground water that is separated from an underlying body of ground water by an unsaturated zone. Its water table is a perched water table. It is held up by a *perching bed* whose permeability is so low that water percolating downward through it is not able to bring water in the underlying unsaturated zone above atmospheric pressure.

Perched ground water may either be *permanent,* where recharge is frequent enough to maintain a saturated zone above the perching bed, or *temporary*, where intermittent recharge is not great or frequent enough to prevent the perched water from disappearing from time to time as a result of drainage over the edge of or through the perching bed.

Ground water, unconfined—Water in an aquifer that has a water table.

Head, static, h (L)—The height above a standard datum of the surface of a column of water (or other liquid) that can be supported by the static pressure at a given point; piezometric head.

The *static head* is the sum of the elevation head, h_e, and the pressure head, h_p; that is, $h = h_e + h_p$ (*see* Head, total). Under conditions to which Darcy's law may be applied, the velocity of ground water is so small that the velocity head, $h_v = v^2/2g$, is negligible. *Head,* when used alone, is understood to mean *static head.*

Head, total, H (L) – In a liquid, the sum of three components: (1) *elevation head*, h_e, which is equal to the elevation of the point above a datum, (2) *pressure head*, h_p, which is the height of a column of static water that can be supported by the static pressure at the point, and (3) *velocity head*, h_v, which is the height the kinetic energy of the liquid is capable of lifting the liquid. Thus, the *total head* can be expressed as:

$$H = h_e + h_p + h_v = z + \frac{1}{g}\int_{p_a}^{p} \frac{dp}{\rho} + \frac{v^2}{2g} \qquad \text{(E-4)}$$

where p_a is atmospheric pressure.

Homogeneity – Uniformity. A material is homogeneous if its hydrologic properties are identical everywhere. Although no known aquifer is homogeneous in detail, models based upon the assumption of homogeneity have been shown empirically to be valuable tools for predicting the approximate relationship between discharge and piezometric head in many aquifers.

Hydraulic gradient – The change in static head per unit of distance in a given direction (dimensionless).

Infiltration – The downward entry of water into the soil.

Infiltration rate (LT^{-1}) – A soil characteristic describing the maximum rate at which water can enter the soil under specified conditions, including the presence of excess water. It has the dimensions of velocity.

Isotropy – That condition in which all significant properties are independent of direction. Although no aquifers are isotropic in detail, models based upon the assumption of isotropy have been shown to be valuable tools for predicting the approximate relationship between discharge and piezometric head in many aquifers.

Intermediate zone – The zone extending from the lower edge of the soil water zone to the upper limit of the capillary fringe. This zone may vary in thickness from zero, when the water table is so high that the capillary fringe extends into the soil water zone, to several hundred feet, when the water table is very deep. Nonmoving water in the intermediate zone is held in place by hydroscopic and capillary forces. Temporarily, excess water moves downward through this zone by gravity drainage.

Interstices – Positions of a rock or soil not occupied by solid matter; pore spaces. Interstices may be divided into two groups: *primary* and *secondary*. Primary or original interstices came into existence when the rocks were formed and can be subdivided into those of sedimentary origin and those of igneous origin.

Secondary interstices developed after the rocks were formed. Examples include joints, fractures, and solution openings formed by plants and animals.

Multilayer aquifer – An aquifer system formed of alternating layers of aquifers and aquitards. Aquifer layers may be considered to be confined, unconfined, or semiconfined.

Porosity, n – In a rock or soil, the property containing interstices or voids. It may be expressed quantitatively as the ratio of the volume of its interstices to its total volume. It may be expressed as a decimal fraction or as a percentage. With respect to the movement of water, only the system of interconnected interstices is significant (*see* Porosity, effective). Porosity is controlled by shape, sorting, and packing arrangements of grains and is independent of grain size (dimensionless).

Porosity, effective, n_e – The amount of pore space available for fluid transmission. It is expressed as a percentage of the total volume occupied by the interconnecting interstices. Effective porosity has been used to mean about the same thing as *specific yield* in most field studies.

Potentiometric surface – A surface that represents the static head; used interchangeably with the term *piezometric surface*. As related to an aquifer, it is defined by the levels to which water will rise in tightly cased wells. Where the head varies appreciably with depth in the aquifer, a potentiometric surface is meaningful only if it describes the static head along a particular specified surface or stratum in that aquifer. The water table is a particular potentiometric surface.

Pressure, static, p ($ML^{-1}T^{-2}$)–The pressure exerted by a fluid. It is the mean normal compressive stress on the surface of a sphere around a given point. The static pressure does not include the *dynamic pressure*, $\rho\ v^2/2$, and is therefore distinguished from the *total pressure*. The velocity of ground water ordinarily is so small that the dynamic pressure is negligible. *Pressure,* when used alone, is understood to mean *static pressure*.

Radius of influence of a well (L)–The distance from the center of the well to the closest point at which the piezometric surface is not lowered when pumping has produced the maximum steady rate of flow.

Saturated zone–That part of the water-bearing material in which all voids are filled with water under pressure greater than atmospheric. In practice, capillary water is often not differentiated. *Phreatic zone* although synonymous with *saturated zone* is less commonly used.

Seepage–The slow movement of gravitational water through the soil or rock.

Seepage velocity (LT^{-1})–The rate of discharge of seepage water through a porous medium per unit area of void space perpendicular to the direction of flow.

Specific discharge, or specific flux, q (LT^{-1})–The rate of discharge of ground water per unit area measured at right angles to the direction of flow. Specific discharge has the dimensions of velocity as follows:

$$q = \frac{Q}{A} \tag{E-5}$$

Soil water zone–The zone extending from the ground surface downward through the major root zone. Its thickness varies with soil type and vegetation. Water in this zone moves downward during infiltration and upward by evaporation and plant transpiration. Temporarily, during a short period of excessive infiltration, the soil in this zone may be almost completely saturated.

Storage, specific, S_s, (L^{-1}) – The volume of water released from or taken into storage per unit volume of the porous medium per unit change in head. In problems of three-dimensional transient flow in a compressible ground water body, it is necessary to consider the amount of water released from or taken into storage per unit volume of the porous medium.

Storage coefficient, S – The volume of water an aquifer releases from or takes into storage per unit surface area of the aquifer per unit change in head. In a confined water body the water derived from storage with decline in head comes from expansion of the water and compression of the aquifer; similarly, water added to storage with a rise in head is accommodated partly by compression of the water and partly by expansion of the aquifer. In an unconfined water body, the amount of water derived from or added to the aquifer by these processes generally is negligible compared to that involved in gravity drainage or filling of pores; hence, in an unconfined water body the storage coefficient is virtually equal to the specific yield.

$$S = S_s H \tag{E-6}$$

where H is the thickness of the aquifer.

Specific yield – The water yielded by gravity drainage as occurs when the water table declines. It is the ratio of the volume of water yielded by gravity to volume of rock. *Specific yield* is equal to *porosity* minus *specific retention.*

Specific retention – The ratio of the volume of water a saturated rock will retain against the pull of gravity to its own bulk volume.

Saturation – The percentage ratio of the volume of water to the volume of void space; also referred to as *moisture content.*

Storage, bank (L^3) – The change in storage in an aquifer resulting from a change in stage of an adjacent surface-water body.

Transmissivity, T (L^2T^{-1}) – The rate of flow of water through a vertical strip of aquifer one unit wide, extending the full saturated thickness of the aquifer, under a unit hydraulic gradient:

$$T = HK \tag{E-7}$$

where H is the aquifer thickness.

Unsaturated zone – The zone between the land surface and the water table. It can be divided into three subzones, in descending order: the *soil water zone*, *intermediate zone*, and *capillary fringe*. *Zone of aeration* is an obsolete term and *vadose zone* is infrequently used.

Unconfined aquifer – An aquifer in which a water table (phreatic or free surface) serves as its upper boundary.

Water table – An imaginary surface in an unconfined water body at which the water pressure is atmospheric. It is essentially the top of the saturated zone.

Index